AF651878

Cyber Livestock Communication and Extension Education

About the Author

Dr. G.R.K. Sharma holds a doctoral degree in Veterinary Extension Education. He is presently an Assistant Professor and Head in the Discipline of Veterinary & Animal Husbandry Extension, College of Veterinary Science, Acharya N.G. Ranga Agricultural University, Tirupati, India. He has to his credit more than 20 publications. His theoretical contributions include a Cyber Livestock Communication Model in Rural India.

Cyber Livestock Communication and Extension Education

G.R.K. Sharma

Assistant Professor & Head
Department of Veterinary & Animal Husbandry Extension
College of Veterinary Science
Tirupati, Andhra Pradesh

CONCEPT PUBLISHING COMPANY, NEW DELHI-110059

Cataloging in Publication Data—DK
Courtesy: D.K. Agencies (P) Ltd. <docinfo@dkagencies.com>

Sharma, G. R. K.
Cyber livestock communication and extension education
G.R.K. Sharma.
p. cm.
Study conducted in Kuppam area of Chittoor District, Andhra Pradesh, India.
Includes bibliographical references (p.)
Includes index.
ISBN 8180693198

1. Internet in agriculture—India—Kuppam. 2. Communication in agriculture—India—Kuppam. 3. Livestock improvement—India—Kuppam. 4. Rural extension—India—Kuppam.
I. Title.

DDC 636.009 548 4 22

All rights reserved. No part of this work may be reproduced, stored in a retrieval system, or transmitted in any form or by any means, electronic, mechanical, photocopying, microfilming, recording or otherwise, without the prior written permission of the copyright owner and the publisher.

ISBN 81-8069-319-8

First Published 2006

© G.R.K. Sharma

Published and Printed by
Ashok Kumar Mittal
Concept Publishing Company
A/15-16, Commercial Block, Mohan Garden
New Delhi-110059 (India)
Phones: 25351460, 25351794
Fax: 091-11-25357103
E-mail: publishing@conceptpub.com

PREFACE

India is passing the era of information and communication technology, which has initiated an interface between computing and communication technologies by rapid devlopment in internet, electronic, telecommunication, computer, multimedia, video-conference and so on. The biggest advantage of services rendered by information and communication technology is that it is more interactive and personalized and can render the information as per the need and requirement of ultimate users. Among the information and communication technologies, internet has ensured the possibility for quick information gathering, processing, transmission, preservation and sharing of information for social, economical and cultural upliftment. The growth of technologies in the field of communication leads to a global spread of knowledge and application of which could leave multi-dimensional impact on all spheres of human activities, more specifically by accelerating the process of information exchange and reducing the cost involved in achieving an ultimate objective at farmers' field. The information and communication technology can render technical services on agriculture and livestock production system, market information, weather information, state and central government policies formed for welfare of farming community. The use of information and communication technologies to assist the farmers in taking right decision at right time in carrying out their farm operation is in nascent stage in India.

Internet is rapidly exanding in developing countries, but most rural communities are not yet able to take advantages of the services available to their urban neighbours. Internet is a medium of communication and perhaps the most flexible medium currently available. Internet is emerging as a tool of improved communication and information access, directly related to social and economic development of rural population and helped to reach the unreached, foster new voices and leaders. It is well known that the Internet has the potential to help the rural poor to leap-frog some of the traditional barriers for development, by improving access to information, expanding their market base, enhancing employment opportunities and making government services much better. The new information and communication technologies like Internet appear to offer a cheaper force of information infrastructure and can be a milestone in improving rural citizen accessibility to information and communication technologies, which give rise to hope that the country will achieve the technology loop to solve the problems linked with development of agriculture and animal husbandry.

India has some inherent problems that are barrier in launching information and communication technologies in agriculture and animal husbandry sector; these problems are large distance and inadequate infrastructure. Further, majority of farmers are in rural area, prossess small holding size, small size of livestock and are not in a strong economic position to afford the cost of information technologies. Despite these problems, the ongoing IT revolution has opened huge opportunities for providing access to information as well as interactive distance learning in rural India. The farmers, who are exposed to the traditional type of information sources may not be able to meet the new challenges in the arising contexts of development and requires supplementary type of information sources, such as Internet. It should be remembered that knowledge is expanding and technologies are advancing at a very faster rate and the farmers has to take the wings in the magnitude of this change. The pilot projects of "Warna wired project" in Kolhapur and Sangli districts and "Babhaleswar project" in Ahmednagar District of Maharashtra, "Info villages" in Pondicherry and "Gyandoot project" in Madhya Pradesh have conclusively demonstrated that, if the information is available at the door step, farmers are willing to pay and make use of technical, marketing and general information. These Internet pilot projects need to be driven, not by technical concerns, but by human knowledge, communication and social relationship concerns.

On the similar lines, the Government of Andhra Pradesh has started "Kuppam project" wherein the information and communication technology tools were extensively used to disseminate outreach information to the end users at the cutting edge and play a vital role in improving important aspects of like, behaviour, resource management, health, production etc. of farmers. In spite of number of activities and projects oriented towards the transfer of information and communication technologies, still the acceptability and utilization of the same was found to be very low in the rural arena, this created the need to probe knowledge utility, practicality and constraints in utilization of information and communication technology by livestock farmers. Development communication has been the subject of interest and the subject of present was well conceived, thought out and analyzed before it took the final shape in the form of book.

I do not consider myself as the most learned guy on Internet communication, still I feel confident that my experience in this field can help the readers and farmers a lot. After conducting the impact study on "Kuppam project" as a part of my doctoral programme, I decided to give expression to all my experiences in the shape of a book. Hence it is an authentic work, which would definitely help the readers. I am sure my readers would learn a lot while reading this book. If you have any problem or query, write to me, your suggestions regarding this book are always welcome.

G.R.K. Sharma

ACKNOWLEDGEMENTS

I take this occasion to express my gratefulness to all the members of the faculty, in particular my guide Dr. H.P.S. Arya, Principal Scientist and Head, Division of Extension Education, Indian Veterinary Research Institute, Izatnagar for his keen diligence, constant motivation, kind cooperation, sagacious guidance and support rendered to me at every stage. I wish to express my indebtedness to Dr. P. Das Deputy Director General (Extension), ICAR, New Delhi and Prof. Devesh Kishore, Head, Division of Education, Research and Training, IGNOU, New Delhi, who gave me research insight and I count their guidance as a distinct privilege to me.

Where emotions are involved words cease to means, lexicon could not have the words to express the affection, blessings, encouragement and sacrifice of my parents (Sri G.V.K. Sharma and Smt. Mahalaxmi) and wife Shivani and daughter Vaishnavi Sharma for their patience and confidence for encouraging me to work hard for the completion of this book. Above all I thank the almighty Lord Balajee who has given me all strength and perseverance to complete this assignment.

G.R.K. Sharma

ABBREVIATIONS

AP	:	Andhra Pradesh
CAO	:	Community Administrative Officer
CICs	:	Community Information Centers
CLC	:	Cyber Livestock Communication
CRO	:	Community Relationship Officer
F	:	Frequency
FAO	:	Food and Agriculture Organization
GDP	:	Gross Domestic Product
GLO	:	Government Liaison Officer
ha.	:	Hectares
ICAR	:	Indian Council of Agricultural Research
ICTs	:	Information Communication Technologies
IT	:	Information Technology
Kms	:	Kilometres
KVK	:	Krishi Vigyan Kendra
n	:	Number
NATP	:	National Agricultural Technology Project
NGOs	:	Non Governmental Organizations
OBC	:	Other Backward Castes
SADH	:	State Animal Husbandry Department
SAUs	:	State Agricultural Universities
SC	:	Scheduled Castes
SD	:	Standard Deviation
SMSs	:	Subject Matter Specialists

ST	:	Scheduled Tribes
VASs	:	Veterinary Assistant Surgeons
VCP	:	Video Casette Player
VCR	:	Video Casette Recorder
VDO	:	Village Development Officer
VLI	:	Veterinary Livestock Inspector
WWW	:	World Wide Web

CONTENTS

LIST OF TABLES

LIST OF FIGURES

INTRODUCTION 1

.....Knowledge and Information are essential for people to successfully respond to the opportunities and challenges of social, economic and technological changes. But to be useful, knowledge and information must be effectively communicated to people

—Kofi Annan

Communication is a conscious effort to share information, ideas and attitudes with others. Access to improved communication is a crucial requirement for sustainable agriculture and rural development. Modern communication technologies, when applied and adapted to conditions in rural areas of developing countries like India can be used for rural communities to improve communication, increase participation, disseminate information and share knowledge and skills for improving efficiency and increase production. Communication networks that enable information to flow to and from rural communities and agricultural organizations, as well as between and among the many intermediary organizations that touch rural communities and farmers directly or indirectly, will strengthen community resources. Communication networking is increasingly becoming the trend in the information world, whether it is done at the interpersonal level or at institutional level and is exerting increasing impetus in information acquisition and utilization.

Livestock have been an integral part of rural livelihood system in India for thousands of years. More than 80 per cent of rural households maintain one or other type of livestock as a source of income and risk absorbing factor in case of crop failure. The share of livestock in Gross Domestic Product (GDP) is 6.3 per cent (Agricultural Statistics, 2001) and is deeply affected by global economic, environmental and marketing forces. Without communication capabilities required to access, analyse and share the information that in turn required to create knowledge, remains at mercy of global market forces. The Internet communication is one tool that can enhance this flow of information to improve production.

Information Technology: Changing Scenario

At the end of twentieth century, people in rural and remote areas of developing countries are facing many unprecedented challenges brought on by the changing global economy, environmental degradation, demographic pressures and food security. To deal with these challenges and to make critical decisions, people at all levels of society must be able to access essential information and communicate, as improved communication and information access are directly related to social and economic development (World Bank, 1995). As we move towards the twenty-first century, terms such as 'global village', 'global information society' and 'information age' are being used to characterize the next century. Nevertheless, there is still a lot of speculation as to what will actually happen. All indicators seem to point to the fact that, the twenty-first century will see an Information Technology (IT) driven world. Consequently, it will become necessary to map out strategies for effectively harnessing IT potentials at national, organizational and individual levels.

The world is entering into an information age and developments in Information and Communication technologies are offering new and cost effective as well as challenging approaches for providing the reach to the population in particular. Technological discoveries of 1980s like television, radio and wide range of audio-visual aids have already brought out considerable changes and made profound impact on rural population. The 1990s have seen the growth, commonly called Information and Communication technology such as telecommunication, teleconferencing, microelec-tronics and Internet.

World had witnessed the growth of Internet users from 20 millions in 1995 to 400 millions in 2000 and the number is expected to reach 1 billion by 2005 and still more than 850 million people in developing countries are excluded from a wide range of Information and Knowledge, with the rural poor in particular remaining isolated from both traditional media and new Information and Communication technologies that would improve their livelihood (Kofi Annan, 1997).

Information and Communication Technologies for Development

Information is critical to the social and economic activities that comprise the development process. Internet, as a means of sharing information, is not simply a connection among people, but a link in the chain of the development process itself (Hudson, 1995). Internet is not a panacea for rural and agricultural development, but it does bring new information resources and can open up new communication channels for rural communities and agricultural organizations. It offers a means for bridging the gaps among professionals, rural people and agricultural producers through the initiation of interaction and dialogue. It can foster new alliances and interpersonal networks together with lateral and cross-sector links among organizations. Most importantly, it can support mechanisms that enable the bottom-up articulation and sharing of information on needs and local knowledge. Primary benefits include increased efficiency in the use of development resources, less duplication of activities, reduced communication costs and global access to information and human resources. None of these benefits are guaranteed by the technology of the Internet. Instead, they are realized when people work together to make it a decentralized and accessible communication tool (Richardson, 1997).

With regard to rural and agricultural development and the Internet, organizations that support the betterment of rural population and improvements in agriculture have important roles to play. For example, the Food and Agricultural Organisation (FAO) is playing an important role in assisting the establishment and growth of Internet services for rural communities and for agricultural development. In cooperation with local agencies, organizations like FAO can help rural communities and agricultural sectors in realizing the benefits of improved communication and access to information.

Importance of Rural Internet Access

The Internet is rapidly expanding in developing countries, but most rural communities are not yet able to take advantages of the services available to their urban neighbours. Internet is a medium of communication and perhaps the most flexible medium currently available. It has the potential to support rural and agricultural development, in the areas of economic development, community development, research/education, small and medium enterprise development and media networks.

Internet is emerging as a tool of improved communication and information access, directly related to social and economic development of rural population and helped to reach the un-reached, foster new voices and leaders. It is well known that, the Internet communication has the power to cut across social and geographical distance and help people find new ways of facilitating the flow of Information and Knowledge and it is recognized as a development communication tool.

Internet has the potential to help the rural poor to leap-frog some of the traditional barriers in development, by improving access to information, expanding their market base, enhancing employment opportunities and making government services much better.

Cyber Extension: Indian Initiatives

In India, a boom in computers began to take place during 1980s and expanded rapidly in 1990s. Growth in demand for personal computers and rising popularity of the Internet were two major forces driving the growth of the domestic IT industry. The sale of personal computers in India crossed the 5 million mark and the rate of adoption of personal computers by households was about one per cent. Keeping this in view, Government of India constituted IT task force with an objective to increase the penetration of personal computers from current level 1:500 to 1:50 (Singhal and Brown, 2000).

In India, Information technology and Internet are playing a significant role in agriculture and rural development. There are over nine lakh Internet connections and 32 lakh net users in the country, by 1995 and there would be 90 lakh Internet connections and 2.7 crore net users, by 2005 (Singhal and Brown, 2000). The ongoing IT revolution has opened huge opportunities for providing access to information as well as interactive distance learning in rural India.

The pilot projects of "Warna wired project" in Kolhapur and Sangli districts and "Babhaleswar project" in Ahmednagar District of Maharashtra, "Info villages" in Pondicherry and "Gyandoot project" in Madhya Pradesh have conclusively demonstrated that, if the information is available at the door step, farmers are willing to pay and make use of technical, marketing and general information. These Internet pilot projects need to be driven, not by technical concerns, but by human knowledge, communication and social relationship concerns. Thus, the intended results of an Internet pilot project ought to relate directly to improvements in social relationships, knowledge access and sharing, enhancement of communication among people and organization and development of rural community resources and agricultural resources.

Need of the Study

On the similar lines of above said initiatives, the Government of Andhra Pradesh has started Kuppam project three years ago in Chittoor district, wherein the Information and Communication Technology (ICT) tools were extensively used to disseminate outreach information to the endusers at the cutting edge. So far no study attempted to evaluate the effectiveness of these initiatives. The study at hand is aimed to provide certain information, viz. how does IT work? which section of the farming community has benefited most? what kind of information do farmers seek at Internet? what information is actually available on Internet? what is the attitude of the farmers towards Internet? how do the farmers exploit the advantages of Internet? what is the feasibility of Internet communication in the backward areas where sizeable population is below the poverty line? These questions require definite answers before expanding Internet communication in India.

The ICTs play a vital role in improving important aspects of life like behaviour, resource management, health, production etc. of livestock farmers. In spite of number of activities and projects oriented towards the transfer of ICTs, still the acceptability and utilization of ICTs were found to be very low in the rural arena. This created the need to probe knowledge, utility, practicability and constraints in utilization of ICTs by livestock farmers.

Keeping the above facts in view, the study entitled "Internet Communication among Livestock Farmers in Chittoor District of Andhra Pradesh" is conceptualised with the following objectives.

Objectives

General Objective

To study and analyse the factors that are determining the effectiveness of internet communication with special reference to livestock development in Kuppam area of Chittoor district in Andhra Pradesh.

Specific Objectives

1. To study the livelihood profile and information seeking behaviour of livestock farmers.
2. To assess the kind of information available through Internet as perceived by the livestock farmers.
3. To measure the effectiveness of Internet communication in terms of knowledge and perceived benefits among the livestock farmers.
4. To study the factors that influence the perceived benefits of the livestock farmers.

Scope of the Study

- The present study is an attempt to explore the Internet communication characteristics among livestock farmers.
- This study has focused on the mediating structures/settings, which can predict the process of effectiveness of Internet communication in the given context.
- The study would contribute to the fast growing knowledge and theoretical advancement of the ICT with the dearth of studies under rural conditions besides contributing to the methodology by developing different factors influencing the Internet communication process.

- The results of this study could be of great use not only to livestock farmers but also to the social scientists, researchers in the field of ICT, government policy makers and all those who are directly/indirectly involved with IT developmental activities in rural India.
- The results stemming from this study will enhance the communication pattern of livestock farmers through better understanding of the current and desired behaviour, as they are the ultimate users of information in the agricultural sector.
- The findings of the present study might be useful for the IT scientists in anticipating feed-back from the clientele to know the modifications required in the technologies for better acceptance and utilization by the farmers.
- The present study will help to know the problems/ constraints faced by the livestock farmers for successful or effective Internet communication. Besides, this study also emphasizes on the reasons for shifting from traditional to Internet communication.

Limitations of the Study

- Social research itself is handicapped with certain limitations and for that matter this study is not an exception. It is the endeavour of one person for the purpose of academic degree. As such, the investigator was confronted with problems such as time, finance, small sample size as well as other resources and facilities.
- As the respondents were not much accustomed to this type of investigation, added to this, the nature of instrument with regards to eliciting personal opinion, attitude and sometimes left the respondents rely heavily on recall and guesses might have involved some kind and degree of

subjectivity and biasness. This is despite the fact that the investigator has tried best precautions and efforts possible to minimize such tendencies through the measures utilized, i.e. reliability as well as other research techniques.

- Since the locale of the investigation was confined to 5 mandals, as such generalisations could be restricted to the areas under study in particular and other areas where similar conditions prevail in general.

Organization and Presentation of the Thesis

The thesis is presented in seven chapters, the first chapter deals with objectives, scope and limitations. The second chapter outlines a brief review of available literature in the light of present investigation. Third chapter covers theoretical orientation and strategic model of the study. Fourth chapter deals with locale of the study, sampling techniques, measurement of variables and statistical tools used for the analysis of data. Information and Communication Technologies: Indian Scenario was presented in fifth chapter. The results and discussion of the investigation are presented in sixth chapter. Last chapter consists of summary, conclusion, implications of the study and suggestions for future research. At the end, literature cited and questionnaire used for the study were appended.

REVIEW OF LITERATURE 2

Comprehensive review of literature provides proper background information to design the research programme, to analyse the data and in interpreting the findings. Thus, an acquaintance with available literature concerned directly or indirectly with the problem under investigation has been felt necessary to develop a comprehensive idea of the present study.

This chapter has been presented under the following heads:

1. Personal and livelihood characteristics of livestock farmers.
2. Information seeking behaviour of livestock farmers.
3. Internet characteristics of livestock farmers.
4. Information source—credibility.
5. Knowledge level and perceived benefits of farmers using Internet services.
6. Relationship between personal, livelihood and Internet characteristics and dependent variables.

Personal and Livelihood Characteristics of Livestock Farmers

Rao (1982) reported that 58.60 per cent of the respondents had family size of 6.24 member and 52.50 per cent of the respondents belonged to joint family.

Charan and Sharma (1989) conducted a survey on socio-economic profile and found that all the small and marginal farmers belonged to lower caste.

Manjunath and Marilingnnavar (1989) in their study observed that 80 per cent of the respondents belonged to big family size, had livestock farming as their main occupation and were in low-income category.

Ramchand *et al.* (1990) reported that majority of the respondents were illiterates and had medium extension contact and mass media exposure with low level of social participation.

Murthy (1990) revealed that increase in contact with extension agency and mass media on the part of farmers, their information needs increased correspondingly.

Manjunatha (1991) reported that majority of the respondents belonged to low economic status and were of backward caste.

Samantha (1993) conducted a study on decision-making abilities and observed that, majority of the respondents were independent followed by heads of the families.

Edwards and Wajcman (1995) observed that technology itself is a product of social relations, so adoption of innovations favours particular social group as men, type of farmers as progressive and large farmers.

Nie *et al.* (2001) observed that dominant group of the society has more access to latest communication technologies in any geographical area in particular.

Information Seeking Behaviour of Livestock Farmers

Bouge (1962) emphasied the superiority of personal communication over mass media for creating motivational effect.

Sawhney (1967) reported that farmers with large holding used personal cosmopolite sources to a large extent than small farmers.

Bhogle (1971) reported that the informal sources seem to be more useful, intelligible and convincing in popularizing the new varieties and practices.

Kulkarni (1973) found that the frequency of visits or contacts with extension agents or other officials had a significant influence on the rate of adoption of new practice in dairying.

Tyagi and Sohal (1984) reported that media exposure demonstrated its influence in increasing knowledge of dairy innovations particularly to the urban respondents. Their study also revealed that extension contact followed by mass media participation were the most effective in increasing the knowledge.

Bhaskaran and Rao (1985) observed that among all the available channels of communication, inter-personal communication channel is widely used in Indian rural communities because through this channel, they enjoy the confidence of fellow farmers to a greater extent.

Ernest (1987) reported that the utilization of mass media sources was found to be positively and significantly correlated with overall adoption level of dairy innovations of landless and medium farmers. However, he further reported that the utilization of mass media sources was found to be positive and non-significantly correlated with overall adoption level of dairy innovations of marginal and large farmers.

Ayaz (1989) revealed that both interpersonal and mass media should be combined together to salvage the best of each. Information should also be relevant to the local situation.

Krikman (1999) observed that friends, progressive farmers and veterinarian were important sources of information for the farmers in the process of accepting a new technology.

Tiamiyu (2000) found that, 53 per cent of the Internet users had changed their information seeking habits, among which decreased frequency to visit library is most important.

Pigato (2001) reported that impersonal-cosmopolite sources like television, media and Internet had dominated in communication of information on any subject and also observed that, veterinarian and progressive farmers were found to be more popular sources of information for scientific animal husbandry practices.

Yadav (2001) reported that extension workers and subject matter specialists were found to be more important sources of information.

Singhal and Rogers (2001) reported that utilization of impersonal cosmopolite sources of information had positive and significant relationship with knowledge level of dairy farmers. He further reported that majority of the livestock owners used animal husbandry department personnel as effective personal-cosmopolite sources of information for selected animal husbandry practices.

Internet Characteristics of Livestock Farmers

Tallero and Gaudette (1995) reported that progressive farmers used Internet more frequently than small and marginal farmers.

Richardson (1996) observed that modern communication technologies, with its enormous advantages like easy accessibility, world-wide information, marginal cost etc., have brought the shift from traditional communication to Internet communication among the rural people.

Shashaani (1997) found that gender differences in Internet access and usage were high and in turn had influenced the knowledge possession.

Khaldoun (1998) found that *none of the Internet users were illiterates* and all of them were educated up to high school level.

Sherman *et al.* (1998) reported that men had significantly higher levels of participation than women in familiarity and Internet activities.

MSSRF (1998) reported that farmers especially small and medium farmers wanted up-to-date information, especially whose wages were partly paid in grains, wanted to know the latest grain prices.

King and Martin (1999) indicated that Internet had been a male domain since it's beginning. The gender gap in Internet use has narrowed in recent years but has not closed entirely.

Sharma (1999) found that much of the rural population preferred and shown inclination towards Internet technology as it accelerates education as well as marketing and income generating opportunities of small and marginal farmers.

Ruth (2000) observed that middle aged and youth were more prone to latest communication technologies and the most productive groups of farmers showed significantly higher scores for Internet use than the less productive group.

Sood (2000) observed that local language had a predominant role in accessing information through Internet and many people used English to access the Internet, as most of the websites were in English.

Chi-cheng Chang (2001) reported that 57 per cent of the Internet users tended to browse Internet for more hours.

Ramanand (2001) reported that farmers had shown keen interest towards seeking technical information followed by government schemes through the latest form of communication technology, i.e. Internet technology.

Das (2002) observed that out of 1949 farmers 141 visited more than 10 times, 248 in between 6-10 times, 1057 in between 2-5 times and 503 only once. He had also observed that 11 per cent of the farmers visited the CICs covering a distance more than 26 kms, 16 per cent in between 11-25 kms, 22 per cent in between 6-10 kms and 41 per cent less than 5 kms.

Kim Deok (2002) reported that 53.9 per cent of Internet users used it connected in cyber cafes, 27.85 per cent at home and 18 per cent in the offices.

Maniar (2002) found that Internet helped not only to become aware about its scope but also to become conversant as well as equipped in utilizing this medium services effectively. Majority of the Internet users had preferred afternoon hours to access Internet, spent about 1-2 hours. He also observed that they were using Internet technology for the last two years and majority of them had changed their information acquisition habits.

Information Source—Credibility

This is one of the most important factors in information communication. The source of communication needs to be reliable, trustworthy, competent and relevant for effective communication.

Aronson (1963) reported that the potency and success or failure of communication depended upon the nature of its source and recipient's perception.

Rogers and Sharma (1971) observed that personal cosmopolite sources of information played an important role in motivating the livestock owners to adopt new animal husbandry practices and they proved to be more effective sources for the purpose.

Sandhu (1973) conducted a study to find out the best method of measuring the credibility of sources by four different methods—ranking, rating, most-least index and paired technique and concluded that most-least index method was efficient for measuring source credibility.

Choole and Rahudkar (1978) revealed that personal cosmopolite sources were more credible by small and marginal farmers. Mass media like television was perceived to be less credible by large farmers.

West (2000) discussed that there was no government or ethical regulations to control majority of Internet's content, thus more credible sources become harder to distinguish from less credible sources. Documents on the web are trustworthy, free from error and complete in nature. The ease of creating, maintaining and changing the information on web has helped the users to a greater degree in providing latest information.

Flanigan and Metzger (2000) indicated that people considered information obtained online to be as credible as television but not as newspaper. Internet sources are trustworthy, unbiased and accurate in nature.

Knowledge Level and Perceived Benefits of Farmers Using Internet Services

Rao (1982) reported that the mean knowledge score was found to be higher among technology adopters than non-adopters.

Sundaraswamy (1993) reported that majority of the livestock farmers possessed medium level knowledge about livestock farming and the profits gained through livestock farming was moderate.

Chambers and Guijt (1995) in their study stated that Information kiosks enable the farmers and rural residents to enrich their knowledge base and share it with one another as well as with decision-makers.

Tallero and Gaudette (1995) reported that improved communication and access to information were directly related to social and economic development.

Krueger (1995) in his study stated that individuals who have computer skills earn about 15 per cent more than other workers with same education and experience but without computer skills.

Lishan Adams (1996) reported that through modern communication technology food security, poverty alleviation and higher level of education could be achieved.

Richardson (1997) observed that modern communication technologies could be used to increase participation, disseminate information and share knowledge and skills, thus, leads to sustainable and income-generating efforts.

Babb (1998) conducted a survey and found that individuals unfamiliar with computers and the Internet had fewer employment opportunities, as the Internet becomes an increasingly important means of job search.

Meier-Dallach (1998) acknowledges that Internet may help to replace routine contacts but does not substitute decisional contacts.

Rios (1999) found that Internet technologies were used for rural communication to increase participation, disseminate information and share knowledge and skills and establishment of cyber cafes at rural areas were income-generating and leads to sustainable and cost-effective efforts.

Uhegbu (2000) stated that distance of Information centers and distance which IT was covered significantly affects the knowledge of the women entrepreneurs.

Das (2002) observed that technical information contribute 64 per cent of the requirement, market information as high as 21 per cent, whether information 6 per cent, besides 9 per cent other types of information.

Cecchini and Scott (2003) emphasized that ICT could help farmers by improving the outlets of their products. Positive approach to latest information and involvement of grassroots intermediaries of the community were the prerequisites for the successful use of ICT by the rural poor.

Relationship Between Personal, Livelihood and Internet Characteristics and Dependent Variables

Hiltz and Johnson (1999) in their study found that individual's status in social and economic strata had a great role to play in their knowledge possession.

Ruth (2000) recommended that the younger generation had shown much interest towards Internet communication and possessed more knowledge pertaining to Internet technology.

Sood (2000) found that Internet acccss through local language had increased the number and time of access. In turn this had shown a positive impact on knowledge level of the Internet users.

Das (2002) emphasized that increased number of visits to CICs and greater utility value of information through Internet had certainly improved the knowledge level of users.

On the basis of the above review of relevant literature, a list of various independent variables was prepared. However, not all of them are included in the present study. Only those who were relevant to the rural conditions and suitable to the objectives of the study were included along with new variables. Opinion of experts in IT was sought to finalize the variables, results of the pre-test and consultations with extension and IT personnel were also considered.

THEORETICAL ORIENTATION 3

In the absence of any known research in the area of IT in agriculture and allied fields like livestock production, this study is the first attempt of its kind. It critically analyzes the effectiveness of Internet communication among livestock farmers. Any study without proper theoritical background tends to be a superficial one. Keeping this in view in this chapter, an effort has been made to lay a basic foundation on a sound theoretical orientation for Internet Communication followed by information age and new ICTs and strategic model.

Communication Theories

Genesis

Three kinds of investigators have made most of the philosophical studies of human communication: mathematicians, social psychologists and linguistic anthropologists. Human communication has been divided into three principle parts: syntactics, semantics and pragmatics. Thesc investigators and parts cut across one another, creating a matrix of three rows and three columns that formulate and answer most questions about human communication and has been evolved with three different communication theories: mathematical, social psychological and linguistic.

Mathematical Theory

The mathematical theory stems from the work of mathematicians and engineers who helped create the modern radio, telephone and television as means of electronic communication. These mathematicians and engineers also created the computers and other instruments of modern automation in communication.

Claude E. Shannon proved to be the key theorist, conceptualizing the theory in the late 1940s that remains central to communication study today. Published, with Warren Weaver, as a Mathematical Theory of Communication (1948), Shannon's theory was born out of his research at Bell Labs. "Shannon's initial goal was simple: to improve the transmission of information over a telegraph or telephone line affected by electrical interference, or noise. The best solution, he decided, was not to improve transmission lines but to package information more efficiently" (Horgan, 1990). Shannon's concept was quickly adopted by researchers in various disciplines and applied to computer science, physics, molecular biology and biotechnology, psychology, linguistics and communications.

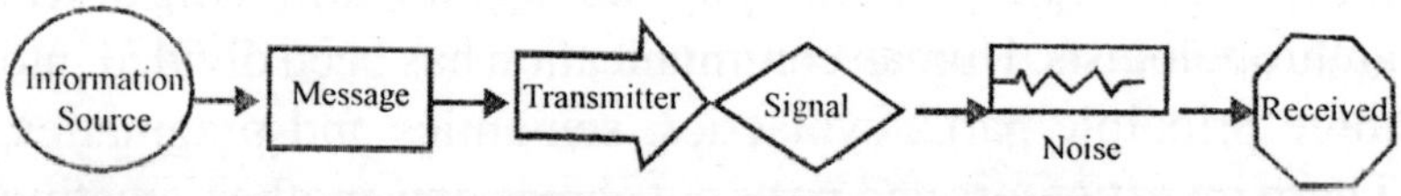

Fig. 3.1: Linear Transmission of Messages

Mathematical Theory and Human Interactions : A Shannon's Model

Shannon clearly designed his theory as a mathematical model that does not take human emotions and experiences into account

(Rogers, 1995). However, communication scholars immediately applied the theory to human interaction. Schramm adapted the model to deal with the concern of "communication, reception, and interpretation of meaningful symbols—processes at the heart of instruction" (Heinich *et al.*, 1996).

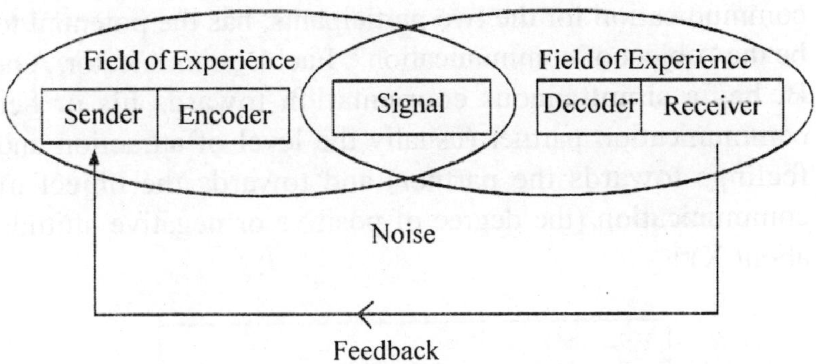

Fig. 3.2: Shannon model of communication

The success of this interaction would then be measured by the feedback the receiver would give to the sender once the message has been transmitted. Feedback, in an instructional setting, may take the form of discussions, observations or tests. Schramm's adaptation provides a communication model that is measurable and adjustable for the production of effective communication in an instructional setting.

Social Psychological Theory

The key orientation for the social psychological theory of communication is in the analysis of human codes and networks during social interactions. Among the social psychologists some have given greater emphasis to the social aspects of communication while others have emphasized the psychological aspects.

Newcomb (1953) introduced the concept of coorientation as the basis of human communication. He developed ABX model as a helpful tool in relational analysis of dyadic pairs. This simple yet insightful model consists of two communicators, A and B, and their "orientation" towards some "object of communication", X. Any subject, behaviour, attitude, belief, event, or object, which is the focus of communication for the two participants, has the potential to be the "object of communication". Each communicator, A or B, has a simultaneous coorientation towards his or her communication partner (usually the level of attraction and feelings towards the partner) and towards the object of communication (the degree of positive or negative attitude about X).

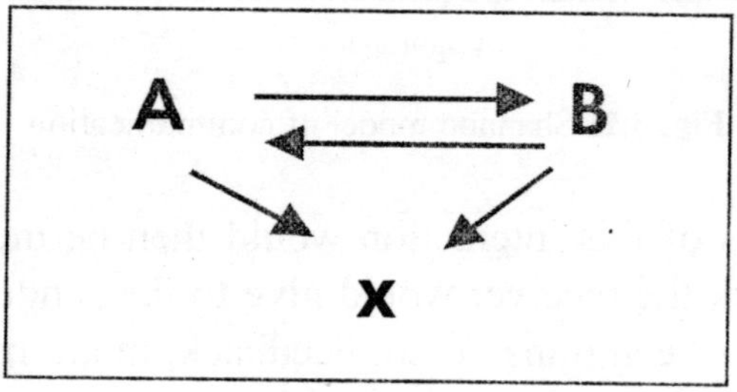

Fig. 3.3: Newcomb's A-B-X or Coorientation Model

Newcomb sees four basic components of this relational system: A's attitude towards X; A's attraction to B; B's attitude towards X and B's attraction to A.

In this model, at the very minimum, any thorough index of a dyadic relationship should include the following two items of information: (1) each person's orientation (that is, his attitude towards the object of communication and his attraction towards his communication partner) and (2) what each person perceives his partner's orientations to be.

Linguistic Theory

This theory is primarily concerned with analysis of speech, but speech is more than just another form of signalling: it is also a form of human behaviour in general. The aim of this theory is to develop a model and a procedure for making a complete, concise and objective description of thc speech of a people.

Greenberg (1966) introduced this theory with the description of verbal behaviour. The linguistic descriptions emphasize patterns and structures; they do this by isolating structural units such as phonemes and morphemes (are basic elements in the sound and grammatical systems of language). Using linguistically based concepts, Greenberg describes not only what the people say or have said, but how they act.

Continuing Applications of Communication Theory

Applications of Shannon's theory are prevalent in everyday life, especially in areas where "computers are a key component, telephone systems, banking, airline reservations, scientific research, weather forecasting and hundreds of other applications of information technology (Rogers, 1995).

Researchers stress that the importance of Shannon's theory cannot be overstated: "It's like saying how much influence the inventor of the alphabet has had on literature" (Horgan, 1992). The linear model presents the measurable variables for all communication processes.

Future applications went far beyond the original focus on channel capacity. For example, Shannon's research helped to develop the idea of check digits, or, error correcting codes, (what Shannon referred to as "correcting device") in a string of binary numbers which is applied in computer technology, such as floppy disks and compact disks.

With the advent of Internet, the ideas of effective communication to the masses moved to the forefront of research in the developing countries. Whether it is in terms of communication for effective instruction, or theory to enhance technological processes, the communication theory that was developed in the 1940s had an immeasurable impact on all forms of communication today, both human and non-human. Theories about mass communication process need to be set forth as systematic sets of prepositions that show in straight forward terms just what is supposed to be related to what in terms of dependent and independent variables. To do this, orders of dependency between propositions can be established. If this is done, the theories can be tested empirically and their validity can be adequately assessed.

Information Age and New ICTs

In this regard efforts have been made to present this section in the following two sub-heads:

Information Age

Homo sapiens survived for over 30,000 years by napping flint into tools for hunting, processing food and building shelters. Metal tools fulfilled the same role for a little more than 10,000 years. The industrial revolution started just over 200 years ago. The information age is barely thirty years old (Heap *et al*., 1995). But the impact of information revolution with its powerful information and communication technologies shrinking the whole universe into a small global village enabling easy and faster sharing of knowledge among the world communities.

Information revolution adds huge new capacities to human intelligence and constitutes a resource which changes the way we work together and the way we live together. Knowledge and Information are basic ingredients of food security and are essential for facilitating rural development and bringing about social and economic change (Pye, 1958 and Munyna, 2000). The relevance of this revolution is supported by Balit *et al*. (1996) who pointed out that the least expensive input for rural development is knowledge.

Information in the field of agriculture and allied sectors, to be of benefit, has to be tailored to local agro-ecological and socio-economic conditions. It has to be backed up by relevant input supply services, credits, market prices and public policies. Synchronization in time and space between knowledge and input delivery systems is essential to impart credibility to the extension message (Swaminathan, 1993).

But, accessing information coming from physically remote rural locations in the developing world was until recently, difficult and costly; it was equally difficult to deliver information to the farmers, extension workers and researchers who live and work in such places.

New Information and Communication Technologies

New information and communication technologies (ICTs) have played a major role in diffusing information to rural communities and have much more potential of getting vast amounts of rural population in a more timely, comprehensive and cost-effective manner. As a great social leveller, information technology can raise cultural barriers, overwhelm economic inequalities and even can put unequal human beings on an equal footing and that makes IT the most potent democratizing tool ever devised.

Among the new ICTs, the Internet with its two powerful communication tools has become quite popular among users. Electronic mail (*e-mail*) is the most commonly used new ICTs and has caused a cultural revolution in the way of individuals and organization interact, in terms of time, cost and distance. The second most significant use of new ICTs is the World Wide

Web(WWW), which enables people to access information on millions of other computers.

With new ICTs, rural communities can acquire the capacity to improve their living conditions and become motivated through training and dialogue with others to a level where they are empowered and they make decisions for their development. Analytically, ICT application in rural and agricultural development as empowering the rural citizens through access to information and knowledge, community development, economic development for agricultural producers, small and medium enterprise development and expand employment opportunities in rural areas and to improve the services to rural citizens.

Strategic Model

On the one hand, we all understand what communication is well enough to talk about it meaningfully, since we all do it from a very young age. On the other hand, personal experience shows that people get very easily confused about the communication that occurs in the real world. Besides this, it is essential that the information availability is demand driven rather than supply driven. The challenge is not only to improve the accessibility of communication technologies to the rural population but also to improve its relevance to local development. The world is rapidly shrinking to a global village and the combination of communication and computer has become so powerful that no sector of human activity can afford to ignore it.

Thus, conceptualizing a livestock communication model and defining anything that can be fit into this model as Cyber Communication was felt necessary to explain the framework for the study of Internet communication. For this Cyber Livestock Communication (CLC) model (Fig. 3.4), has been conceived from the Internet and the computer revolution itself.

The CLC model centres around information received by livestock owners from different sources like State Agricultural Universities (SAUs), veterinary research institutes under ICAR, State Animal Husbandry Departments (SAHDs) and KVKs, NGOs and other input agencies in addition to print and electronic media. Through creation of databases to retrieve information, full text retrieval systems and hypertexts with multimedia features by integrating audio/video and still images, value addition to the received information in a highly interactive form takes place. The data can be varied covering animal health, breeding, feeding and management aspects, disease outbreaks, marketing information, price of animals, milk, meat, eggs, hides, fish in the adjoining areas or district headquarters and input availability should be collected to create dynamic bulletin systems for use within the area. The databases can be rendered highly interactive and farm-friendly interfaces. The use of appropriate icons recognizable by the farm families and touch screen technology can help to overcome the language/literacy barrier. The possibilities of easy access and acquisition of new skills to interact with computers will provide an enabling environment for the farm families to enhance livestock productivity on a sustainable basis. Inclusion of general developmental information dynamically augments the relevance and usability of CLC model.

The CLC can be accessed in at least two ways. One at the district level direct access by the user visiting the CLC system where it is located and the other is access through the Community Information Centres (CICs) located at village/ mandal/ block level. At the end users' level, CICs will play the role of information dissemination point. Through CICs villagers can exchange information with the neighbouring villages or at elsewhere. These CICs communicates about the information on livestock outreach and in turn delivers it to the end users. The information can also be simplified by easy to follow text and audio-visual

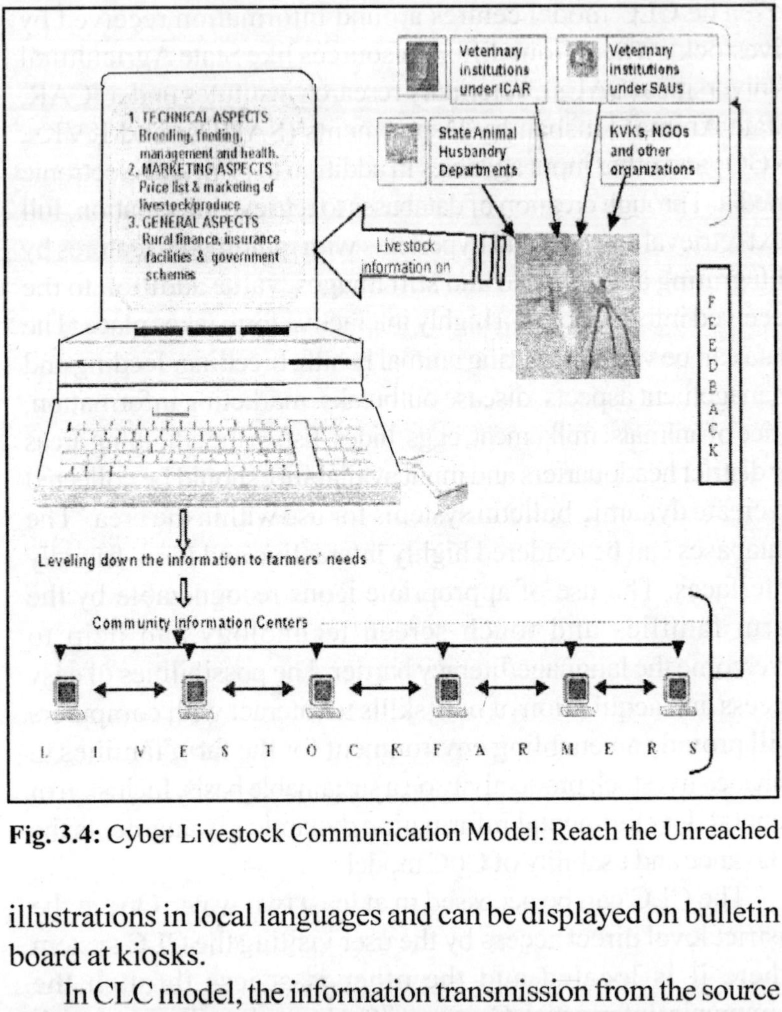

Fig. 3.4: Cyber Livestock Communication Model: Reach the Unreached

illustrations in local languages and can be displayed on bulletin board at kiosks.

In CLC model, the information transmission from the source to CIC will be under public domain. While from CICs to livestock farmers, the information is available under private domain.

Potential Advantages of CLC Model over Traditional Communication

The potential advantages of Internet communication (CLC model) over traditional communication are discussed below:

Continuously Available

Internet communication knows no time restraints. One of the key attributes of an online information service is that it is available all the time, 24 hours a day and 365 days a year. It can be accessed at any convenient time of the individual livestock owner.

Information Reach

Internet contains huge amounts of information and it appeals to the curious subject matter specialists and analytical farmers. It will allow them to search and locate the information they need quickly. The rural technologies can be made available on CD-ROMs for quicker dissemination among the specialists as well as farmers. The Internet has the ability to present and access information of desired depth.

Offer Instant International Reach

Online networks have created an instant global village. Internet communication will eliminate the time and distance barrier that get in the way of knowing the latest information on any particular problem from any part of the world. One can reach any university, national, state level training and research institutions and other research station and discuss his/her problems with the best scientists/experts in the field. The Internet can therefore be seen as a huge reference library.

Cut Steps from Traditional Process

CLC model will remove a number of steps altogether from the traditional extension process. The scientists can directly post the information on the Internet, which will be directly available to extension functionaries at district, subdivision, block and village level. All the concerned will get the information immediately and queries/clarifications will also be addressed equally fast, without involving a chain of extension machinery.

Receiver-Oriented Approach

Strong competition and the realization that the receiver is king leads to the communication process. Receiver is either individual or mass audience, segmented in Internet communication.

Activism

Unlike in traditional communication, receiver is an active participant in Internet communication.

Individualism

Unlike in traditional communication, receiver is not time-bound and can access the information whenever he/she has leisure hours under CLC. Similarly, receiver can even access the information on any subject and to any extent.

Message

In Internet communication, the receiver can choose the message or information from the global library and can access the same to the deepest extent.

Feedback

In Internet communication, the feedback is immediate and detailed in nature and emphasizes more about individual behaviour.

Save Money, Time and Effort

Scientists will prepare electronic version of messages themselves. These versions don't have to be printed and posted. This will save money and time. Cyber messages will be updated online and that saves time too. Internet communication can provide more in-depth analysis and can also provide detailed on-farm research results to the curious users/farmers.

There are massive programmes to improve cyber communication and public computer networks which can be used by information systems in livestock production, research and development. The stakeholders in IT developing their own information network separately can be used for livestock research development. But a combined approach of all IT stakeholders will lead the way. Similarly, ground level development of information systems will be required to make efficient use of this information.

MATERIALS AND METHODS 4

The methodology was devised in order to show vividly and logically the sequence of actions of the researcher by which the investigation reaches its structural finality. This chapter is intended to present the research methodology followed in this study and was presented under the following sub-heads:

1. Research design.
2. Locale of the study.
3. Sample and sampling procedure.
4. Variables and their measurement.
5. Devices and methods used for data collection.
6. Statistical tools and analytical procedures.

Research Design

The study had followed the Ex-post-facto research design. Kerlinger (1983) defined Ex-post-facto design as any systematic empirical inquiry in which the independent variables have not been directly manipulated, because they have already been occurred or they are inherently not manipulated. He further stated that Ex-post-facto studies can be devised to deduce theories, identify behaviour phenomenon and explore the conditions under which phenomenon occurs.

Locale of the Study

Selection of the State

The state of Andhra Pradesh was selected purposively for the study. Since, Andhra Pradesh state depicts both the prevailing situation of ICTs in our country as well as the competitive global marketing scenario and the investigator hails from the same state and was familiar with local language, which would help to establish quick rapport with respondents.

Selection of the Study Area

Chittoor district was selected purposively for the study. Out of 66 mandals in the district, five mandals, namely Kuppam, V. Kota, Shanthipuram, Ramkuppam and Gudipalle comprising 102 villages were selected purposively for the study. Since,

- — The selected area seems to become India's most Internet enabled rural region, with the state government and Hewlett Packard (HP) launching the world's first i-community.
- — The selected mandals were considered to be the first and only mandals in introducing Chief Minister's Empowerment Yojana (CMEY) program to set up a chain of Community Information Centres (CICs).
- — As the CICs were setup at mandal headquarters, all the villages coming under the jurisdiction of these five mandals were included in the study.

Sample and Sampling Procedure

A list of livestock farmers who visited CICs and were using Internet services to access latest information regarding technical, marketing and general aspects was prepared. Based on the number of computer systems installed in each CIC area (which were installed on the basis of population size and literacy rate of the respective mandals (Appendix-I), respondents were selected through proportionate random sampling to ensure both proportionativeness and representativeness of livestock farmers in the selected area. As such, 80 farmers from Kuppam,

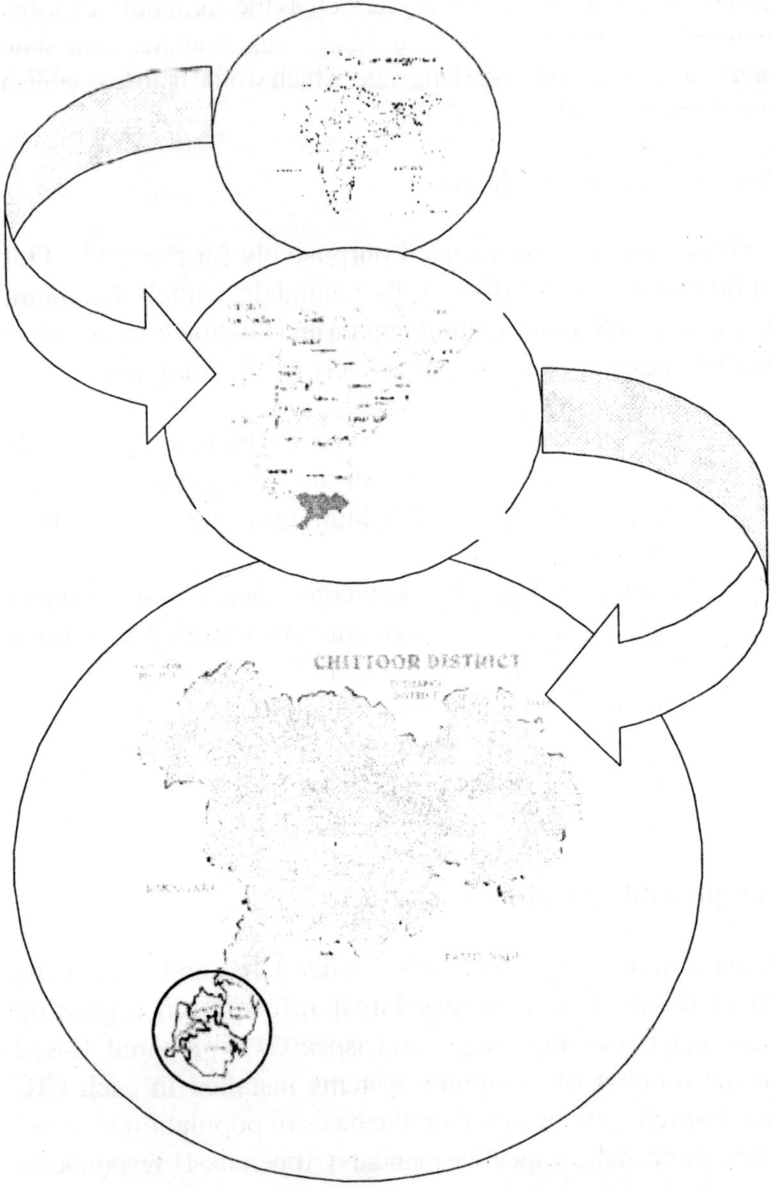

Fig. 4.1: Map shows the selected mandals in Chittoor district of Andhra Pradesh

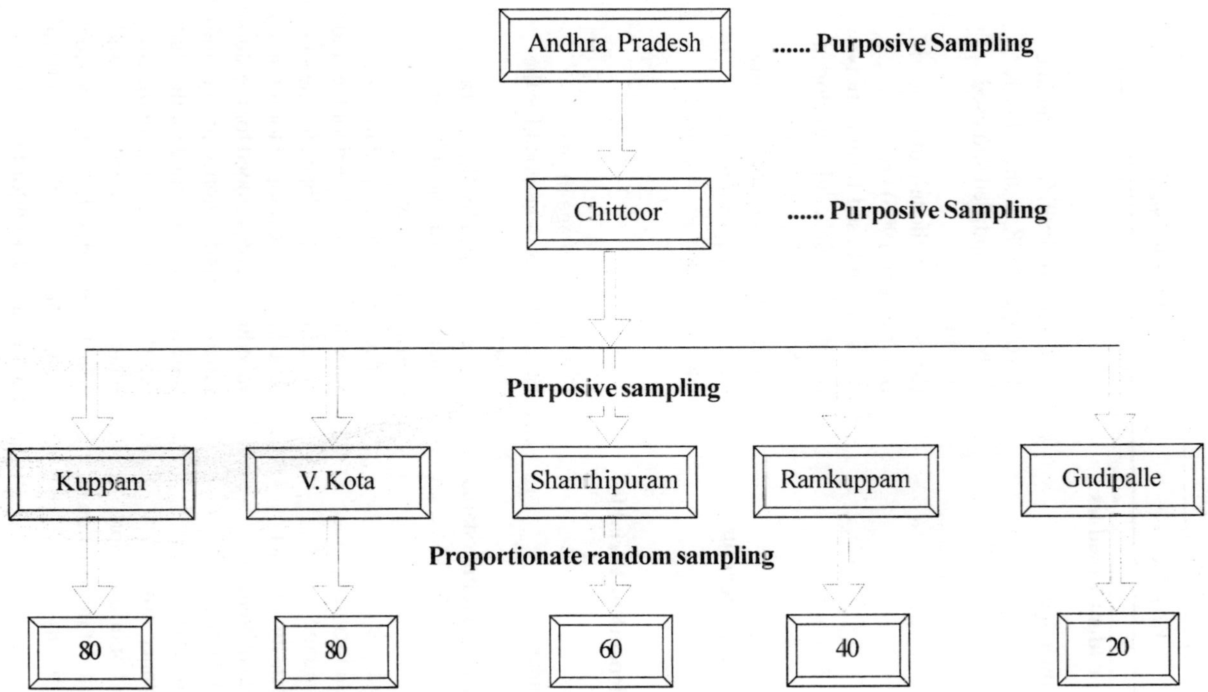

A total of 280 respondents belongs to 102 villages comprising 5 mandals were included in the study to form the sample size

Fig. 4.2: Sampling Procedure Adopted for the Study

Table 4.1: Variables and their measurements

S.No.	Variables		Measurement
I.	**Dependent Variables**		
1.	Knowledge	....	(a) Cattle and buffalo : Schedule developed by Rao and Kherde (1990) was modified and used for the study (b) Sheep : Schedule developed by Goswami (1996) was modified and used for the study (c) Poultry : Schedule developed by Narmatha *et al.* (1995) was modified and used for the study
2.	Perceived benefits	....	Open-ended questionnaire was developed for the study
II.	**Independent Variables**		
1.	Age	....	Chronological age in years
2.	Gender	....	Schedule was developed for the study
3.	Socio-economic status	....	Scale developed by Venkataramaiah (1990) was modified and used for the study
4.	Herd size	....	Schedule was developed for the study
5.	Total Income	....	Schedule was developed for the study
6.	Language	....	Schedule was developed for the study
7.	Extension contact	....	Schedule was developed for the study
8.	Inquisitiveness	....	Schedule was developed for the study
9.	Time of use	....	Schedule was developed for the study
10.	Frequency of use	....	Schedule was developed for the study
11.	Distance covered	....	Schedule was developed for the study
12.	Kind of information	....	Schedule was developed for the study
13.	Utility of information	....	Schedule was developed for the study
14.	Credibility	....	Scale developed by West (2000) was modified and used for the study

80 from V.Kota, 60 from Shanthipuram, 40 from Ramkuppam and 20 farmers from Gudipalle, were selected. Thus, a total of 280 respondents were selected for the study (Fig 4.2).

Variables and Their Measurement

Based on the theoretical orientation, available literature and opinion of experts in the field of ICT, the following variables were selected for the study.

Measurement of Dependent Variables

In the present study the effective Internet communication has been operationalised as the knowledge gained and perceived benefits by the livestock farmers through Internet.

Knowledge

Knowledge was operationally defined for the purpose of investigation, as the information possessed by the livestock farmers about technical, marketing and general aspects of livestock farming.

For measuring knowledge of cattle and buffalo, sheep and poultry keeping farmers, knowledge tests developed by Rao and Kherde (1990), Goswami (1996) and Narmatha *et al.* (1995), respectively were employed with suitable modifications (*See* Appendix IV). The reliability of the tests were established with test-retest (r = 0.84, 0.79 and 0.71), method respectively. The validity of the tests, were obtained through content validity by self-evaluation and judges opinion.

To know the knowledge exclusively gained through internet was measured and presented. To know the relationship between independent variables and knowledge relational analysis was calculated.

Scoring

The knowledge test was executed separately to the cattle and buffalo, sheep and poultry farmers consisted of 60, 39 and 44 items respectively. Test items were of objective type and were administered to the respondents. The response was recorded either as 'Correct' or 'Incorrect' and a score of '1' and '0' were given respectively.

Knowledge scores for the items, viz. breeding, feeding, management, health care, vaccination, fodder production, preparation of milk products, project proposal writing, marketing and general aspects were calculated separately by adopting the following formulae and summed up to arrive at the total knowledge score.

$$\text{Knowledge score} = \frac{\text{Total correct score obtained}}{\text{Maximum correct score}}$$

$$\underset{\text{(Internet)}}{\text{Knowledge score}} = \frac{\text{Total score obtained through Internet}}{\text{Maximum correct score}}$$

$$\underset{\text{(other sources)}}{\text{Knowledge score}} = \frac{\text{Total score obtained through other sources}}{\text{Maximum correct score}}$$

$$\text{Knowledge score} = \underset{\text{(Internet)}}{\text{Knowledge score}} + \underset{\text{(other sources)}}{\text{Knowledge score}}$$

Further one way analysis of variance (ANOVA) was employed to see the significant difference if any in knowledge gain among small, medium and large categories of farmers as well as cattle, buffalo, sheep and poultry farmer and farmers using internet at kiosks and home.

Perceived Benefits

Perceived benefit was operationally defined for the purpose of investigation, as the perceptions about the benefits attained by Internet using livestock farmers through Internet communication.

Open-ended questionnaire were employed seperately to know the percieved benefits rendered by internet communication among different livestock owners. The frequency and percentages were calculated and rank orders were given.

To know the difference in perception of benefit rendered by internet between different category of livestock farmers, the correlation coefficient was calculated and presented. Similarly to know the correlation if any between place of access and perceived benefit, rank order correlation coefficient was computed and presented. To know the relation between the different independent variable and perceived benefits relational analysis was calculated.

Further, the correlation if any between computer literacy and level of information intensity and perceived benefit was calculated and presented.

Measurement of Independent Variables

The independent variables for the study were selected after a thorough review of available literature on the topic and in consultation with experts. The variables which have been found to have some relevance were only included based on the expert judgements (Appendix III). The selected profile characteristics, of the livestock farmers, their operational definitions, measurements and scoring were detailed as follows :

Personal Characteristics of Livestock Farmers

Age: Age was operationally defined as the number of years the individual completed chronologically. Kccping mcan and standard deviation as a check, the livestock farmers were grouped into following three categories:

(a)	Young	=	Below	(Mean - SD)
(b)	Middle	=	Between	(Mean ± SD)
(c)	Old	=	Above	(Mean + SD)

Gender: Gender was operationally defined as the sex of the respondent as male or female. The responses of the individuals were expressed in terms of frequencies and percentages.

Socio-Economic Status

Socio-economic status scale developed by Venkataramaiah (1990) was used with modifications (*See* Appendix IV) to suit the present study. The scale comprised religion, caste, self-education, family educational status, marital status, family type, family size, material possession, communicational aids, self-occupation, family members' occupation, type of house and position in the family items. The scores of all these items were summed up to arrive at total score of socio-economic status of the respondents. Based on mean and standard deviation the respondents were grouped into three categories. However, the frequency and percentage for each sub-component of the socio-economic scale were also computed and presented separately.

(a)	Low	=	Below	(Mean - SD)
(b)	Medium	=	Between	(Mean ± SD)
(c)	High	=	Above	(Mean + SD)

Religion: The religion of a person was conceptualized as the community stratum, one acquires at the time of birth. Scoring was given as below :

S. No.	*Category*	*Score*
i.	Hindu	3
ii.	Muslim	2
iii.	Christian	1

Caste: The caste of a person was conceptualized as the social stratum, one acquires at the time of birth. SC and ST include Yamadi, Yerukula, Lambadi, Dodekula, Bogam, Segali, Mala, Madiga, Thoti etc. Whereas Backward includes Mangali, Chakali, Bestha, Uppara, Idiga etc. While Other Backwards include Kamsali, Kumari, Silpi, Vadrangi etc., and General embodies Brahmin, Reddy, Kamma, Velma, Kshatriya etc. Scoring was given as below :

S. No.	*Category*	*Score*
i.	SC and ST	1
ii.	Backward	2
iii.	Other Backward	3
iv.	General	4

Self Education: Education was operationally defined as number of years the respondent attended teaching institution. The respondents were grouped in seven categories and the scoring for different categories were given below:

S. No.	*Category*	*Score*
i.	No. Schooling	0
ii.	Functionary	1
iii.	Primary school	2
iv.	Middle School	3
v.	High School	4
vi.	Intermediate	5
vii.	Graduate and above	6

Family Educational Status: Family educational status was operationally defined as the overall educational level of all the family members. The family educational status was calculated by given formula:

Family Educational Status = Total educational score/Effective family size

Where,

Total educational score = Sum of educational score of each individual of the effective family.

Effective family size = Total number of eligible family members, i.e. excluding children of below 6 years of age,

Marital status: This was operationalised as the present status of the respondent as unmarried or married. The respondents were grouped into unmarried and married groups and frequency and percentages were calculated.

Family Type: Joint family represents family with more than a pair of couple with married children living together and Nuclear family represents family with single pair of couple with unmarried children living together.

Based on the information obtained from the respondents on the type of family, the respondents were grouped into joint family and nuclear family with '2' and '1' scores respectively.

Family Size: Family size was considered in acceptance with the number of members within a family and was categorized based on the number of family members, the respondent possessed. Scoring for the different categories as given below:

S. No.	*Category*	*Score*
i.	Small (1-3 members)	1
ii.	Medium (4-6 members)	2
iii.	Large (7-9 members)	3
iv.	Very large (above 9)	4

Material Possession: Material possession was considered as the household and farm related items, equipments and implements owned by the respondents. The scoring was done in accordance with the scale developed by Venkataramaiah (1990).

S.No.	*Category*	*Score*
i.	Furniture	1
ii.	Bullock cart	2
iii.	Farm implements	3
iv.	Pump set	4
v.	Tractor	5
vi.	Bicycle	1
vii.	Mobike	4
viii.	Car	5

Categorization: Based on the total scores obtained by the respondents, were grouped into three categories, keeping mean and standard deviation as a check.

(a)	Low	=	Below	(Mean - SD)
(b)	Medium	=	Between	(Mean ± SD)
(c)	High	=	Above	(Mean + SD)

Communication Aids: Communication aid was considered as the access to various audio-visual aids and communication net-works of the respondents. The scoring was done in accordance with the scale developed by Venkatara-maiah (1990).

S. No.	*Category*	*Score*
i.	Radio set	1
ii.	Television set	2
iii.	V.C.R.	3
iv.	V.C.D.	3
v.	Tape recorder	1
vi	Telephone	4
vii	Computer with Internet	5

Self-occupation: Occupation, it was decided based on social status and behaviour of respondents. The respondents were categorized based on frequency and percentage, keeping their main occupation, i.e. livestock farming or others, viz. agriculture, agriculture labourer, service etc. into consideration.

Family Members' Occupation: Occupation, it was decided based on social status and behaviour of family members. The respondents were categorized based on frequency and percentage, keeping their main occupation, i.e. livestock farming or others, viz. agriculture, agriculture labourer, service etc. into consideration.

Type of House: The type of house where the farmer was living and the material that was used in the construction were consi-dered. Scoring was conside-red as given below:

S. No.	*Category*	*Score*
i.	Thached shed	1
ii.	Mudwalls and thatched	2
iii.	Brickwalls and titled	3
iv.	Concrete house	4
v.	Concrete double	1
vi.	Telephone	4
vii.	Computer with Internet	5

Position in the Family: It was operationalised as respondent's position in the family, which helps in indicating his level of participation in the decision making process.

S. No.	*Category*	*Score*
i.	Head	3
ii.	Independent	2
iii.	Dependent	1

Livelihood Characteristics of Livestock Farmers

Land holding : The land holding was operationalized as the details of land possessed by the respondents under different types were converted into standard acres, as one acre of wet land is equal to 2.5 acres of dry land. Based on the number of standard acres as possessed, by the respondents were categorized into six categories, namely landless, marginal, small, low medium, high medium and large. Scoring for different categories were given below:

S. No.	*Category*	*Score*
i.	Landless	0
ii.	Marginal	1
iii.	Small	2
iv.	Low medium	3
v.	High medium	4
vi.	Large	5

Herd Size: Herd size was operationalised as the animal wealth possessed by the respondent such as cattle, buffalo, sheep and poultry belongs to both indigenous and improved breeds. According to herd sizes (species-wise) possessed by the respondents and were grouped into three categories by keeping mean and standard deviation as a check.

(a)	Small	=	Below	(Mean - SD)
(b)	Medium	=	Between	(Mean ± SD)
(c)	Large	=	Above	(Mean + SD)

Rearing Pattern : It was operationalized as the pattern/ system followed by the respondents for rearing of their livestock.

Range System: Animals are kept loose throughout the day and night except at milking time.

Semi-Intensive System: Animals are kept in an open paddock, which is provided with shelter along one side where the animals can retire and provided with a separate milking barn.

Intensive System: Animals are confined together on a platform and secured at neck. The animals were fed, milked as well as treated in this barn.

Deep Litter System: Poultry birds are kept in large pens up to 250 each, on floor covered with litter like straw or saw dust up to depth of 8-12 inches.

Cage system: Poultry birds are confined to a cage just large enough to permit very limited movement and allow the bird to stand and sit comfortably.

Feeding Pattern : It was operationalized as the type of feeding followed among the respondents for their livestock. The respondents were categorized based on their feeding pattern followed, viz. only roughages, roughages along with concentrates, only grazing and grazing along with supplementation and were categorized based on the frequency and percentage.

Total income. Income was operationally defined as the income which is gained by the individual through livestock farming, agriculture, labourer and service.

Scoring: To arrive at the income level of each respondent, the income gained through cattle, buffalo, sheep, poultry, agriculture, agricultural labour and service were summed up separately; the respondents were categorized into following three groups:

(a)	Group-1	=	Below	1 lakh.
(b)	Group-2	=	Between	1 to 3 lakhs.
(c)	Group-3	=	Above	3 lakhs.

Information Seeking Behaviour of Livestock Farmers

Sources of information : Refers to different channels adopted by livestock farmers to aquire scientific or other information through research-extension linkages as well as other media/ methods. It was operationalized interms of personal-localite, personal-cosmopolite and impersonal-cosmopolite. The respondents were categorized based on frequency and percentage.

Extension Contact: Extension contact was operationally defined as the degree to which an individual maintained contacts with personnel of extension, animal husbandry department, university and CICs members like Gram sevak, V.D.O., V.L.I., V.A.S., University personnel, Agricultural officer, C.A.O., G.L.O. and C.R.O. to get information and technical assistance regarding livestock farming.

Scoring: The frequency of contact of respondents with the personnel of extension, department, university and CICs. were measured on the basis of five-point continuum, viz. weekly once, fortnightly once, monthly once, once in six months and never with scores of 5, 4, 3, 2 and 1 respectively. Based on the total score obtained by the respondents, they were grouped into three categories by keeping mean and standard deviation as a check.

(a)	Low	=	Below	(Mean - SD)
(b)	Medium	=	Between	(Mean ± SD)
(c)	High	=	Above	(Mean + SD)

Subscription of Newpapers/Magazine

It was conceptualized as the subscription of newspapers or livestock production related magazines by the livestock farmers inorder to acquire latest information. The respondents were categorized based on frequency and percentage.

Inquisitiveness

Inquisitiveness was operationally defined as the degree to which a farmer has shown inclination towards the latest forms of information acquisition that might take him more accessible to valuable information. Inquisitiveness was conceptualized in this investigation in terms of personal-localite sources, personal-cosmopolite sources, impersonal-cosmopolite sources, extension contact and subscription to various newspapers or livestock production related magazines. Thus the scores of above items were summed up to arrive at the total score of inquisitiveness of the respondents. Based on mean and standard deviation the respondents were grouped into three categories.

(a)	Low	=	Below	(Mean - SD)
(b)	Medium	=	Between	(Mean ± SD)
(c)	High	=	Above	(Mean + SD)

Internet Characteristics of Livestock Farmers

Frequency of Use

For the present study, frequency of use refers to the number of times to which an individual visited the kiosks or utilized the Internet services. The frequency of use of Internet services by the respondents were measured as daily once, weekly once, weekly 2-3 times and monthly once with the scores of 4, 3, 2 and 1 respectively.

Distance Covered: For the present study, distance covered refers to the distance travelled by an individual to reach the kiosks for Internet browsing. This variable was measured as less than 2 kms, 2-5 kms, 6-10 kms and 10-15 kms with the scores of 1, 2, 3 and 4 respectively. Responses of all the respondents were summed up and expressed in terms of frequencies and percentages.

Time of Use: It was operationalised in this investigation in terms of time preference, duration spent and years of usage with Internet either at kiosks or home. This variable was measured by following three items:

Time Preference: According to time preference, viz. morning, afternoon and evening revealed by the respondents, responses of all the respondents were summed up and expressed in terms of frequencies and percentages.

Duration: According to duration, viz. 0-1 hr., 1-2 hrs and 2-3 hrs. with the scores 1, 2 and 3, respectively responses were obtained. Based on the total score obtained by the respondents for the duration spent on Internet browsing, they were grouped into three categories by keeping mean and standard deviation as a check.

(a)	Less	=	Below	(Mean - SD)
(b)	Medium	=	Between	(Mean ± SD)
(c)	More	=	Above	(Mean + SD)

Years of Usage: For the present study, years of usage refers to the number of years got acquainted with the Internet. The years of usage of Internet services by the respondents were measured as less than 1 year, 1-2 years and 2-3 years with the scores of 1, 2 and 3, respectively.

Language: It was conceptualized as the lexemes with concordant phonemes are clustered together within a head-word sememe, without postulating any *a priori* assumptions about the ancestry of the phonemic version of the sememe. In other words, for the speakers of myriad tongues to function within a single administrative unit requires some medium of common communication and the choice of this tongue is known as language. Based on the extent of languages known, respondents were categorized into five categories, i.e. don't know, understand, understand and speak, understand, speak and read and understand, speak, read and write. Data obtained through responses was expressed in terms of frequency and percentage. The scoring was given below:

S. No.	*Category*	*Score*
i.	Don't know	0
ii.	Understand	1
iii.	Understand and speak	2
iv.	Understand speak and read	3
v.	Understand, speak read and write	4

Kind of Information: Kind of information was operationally defined as the information available for the livestock farmers through Internet. For the present study, this variable was operationalised as the perceptions of farmer and investigator towards the availability of information through Internet sources.

The available information in the websites as mentioned by the respondents was checkedout by the investigator. The number of respondents who accessed the information species-wise were workedout, as their average frequency formed the observed frequencies for c^2 analysis and the expected frequencies for c^2

analysis were the number of respondents accessing the information for each species from different websites.

Scoring: Three types of information, viz. technical, marketing and general information, together comprising 20 information items about which the information was available in Internet, were included in the schedule and administered to the respondents. Livestock farmers were asked to indicate the different items of information, which they have perceived either available or not available. Responses of all the respondents were summed up and expressed in terms of percentages.

Utility of Information: Utility of information was operationally defined as the degree of utilization of assistance for livestock farming by the respondents rendered through Internet. For the present study, this variable was analyzed by taking the number of respondents who perceived that the information pertaining to different subject areas were available through Internet.

Scoring: A schedule was developed comprising very useful, useful, somewhat useful and very little useful items with 4, 3, 2 and 1, respectively, to quantify this variable. Data obtained through responses was enumerated and expressed in terms of frequencies and percentages.

Based on the extent of acquaintance with Internet, the respondents might have undergone some changes in their information acquisition habits. Schedule was developed to measure this variable; the responses were recorded in the form of 'Yes' or 'No'. If 'Yes' the extent of changes in information acquisition habits as perceived by the respondents were enumerated and expressed in terms of percentages.

Credibility: In the present investigation, source credibility was operationally defined as the degree to which a communication source is perceived as trustworthy, accurate, complete, latest and competent by the farmers. Credibility of the source of information was measured by using scale developed by West (2000) with modifications (*See* Appendix IV). The original scale contains 10 statements, however, only five statements were found to be relevant to the present study setting. Hence they were adopted after establishing reliability (r = 0.79) and content validity.

S.No.	*Statement(s)*	*Degree of Agreement*				
		SA	A	UD	DA	SDA
1.	Information available through Internet is trustworthy in nature	5	4	3	2	1
2.	Information available through Internet is free from error and correct in nature	5	4	3	2	1
3.	Information available through Internet is complete and adequate in nature	5	4	3	2	1
4.	The Internet sources informs to a greater extent	5	4	3	2	1
5.	The Internet sources maintains and provides latest and timely information	5	4	3	2	1

SA: Strongly Agree; A: Agree; UD: Undecided; DA: Disagree; SDA: Strongly Disagree

Scoring: The scale included five statements and the response of each statement was measured over a five-point continuum and the scores of strongly agree and agree were clubbed and strongly disagree and disagree were clubbed under 'Agreement' and 'Disagreement' respectively. The numbers of undecided respondents were subtracted from the sample size.

Constraint Analysis : The constraints faced by the farmers in internet communication and suggestions offered by them to overcome those constraints were elicited through open ended questionnaire. The responses were tabulated based on frequency and percentages and rank order were given.

Devices and Methods Used for Data Collection

Development of Interview Schedule

The interview schedule was prepared in consultation with the experts in the field of extension and Information Technology and keeping in view the objectives of the study the schedule was prepared in a simple language (at farmer's level of understanding). Before finalizing, the schedule was pre-tested under similar conditions. Based on the experience gained in pre-testing the schedule was modified wherever necessary. The interview schedule was translated into local language (Telugu) so as to make the respondents to understand the questions properly and furnish reliable responses.

The interview schedule consists of three sections. Section-I consists personal, socio-economic and livelihood characteristics of livestock farmers, Section-II consists information and Internet characteristics of livestock farmers and Section-III consists credibility, perceived benefits, Knowledge test (species-wise), problems faced and suggestions given by the livestock farmers.

Establishing Necessary Rapport

A pilot visit was made to the study area to ascertain the feasibility of the study and to establish the much needed rapport for conducting field investigation and also clearing out apprehensions if any, from the minds of respondents. Informal farm and home visits were made by the investigator to gain the confidence and create enthusiasm among the respondents. Further, it was made clear to respondents that, the study was purely academic in nature. Later, all the respondents were interviewed with the help of pre-tested schedule personally by the investigator either at farm, residence or kiosks.

Preparation of Report

The data thus collected was coded, tabulated, analyzed and presented in tables for easy understanding and to make the findings meaningful. The findings emerged out of the data were suitably interpreted and necessary conclusions were drawn.

Statistical Tools and Analytical Procedures

Statistical tools used for analysis of data includes frequency, percentages, arithmetic mean, standard deviation, correlation coefficient, analysis of variance, rank order correlation, coefficient of concordance, chi-square test, two-sample 'z' test, two-sample 't' test (Snedor and Cochron, 1994 and Chandel, 2002).

INFORMATION AND COMMUNICATION TECHNOLOGIES
Indian Senario 5

Knowledge rather than the intensive use of resources is now the key to the future of food security in the developing world. In coming years, sustainable agriculture will rely on knowledge and skills to create more income, jobs and food. The emerging Information and Communication Technologies (ICTs) have a significant role to play. ICT is a key enabler of globalization, it allows for the efficient and cost effective flow of information, products, people, capital and ideas across national and regional boundaries. Keeping this in view, an effort was made to get familiar with the initiation and existence of ICT under Indian conditions. This chapter has been presented in the following sections:

1. Genesis of personal computers and Internet.
2. Digital divides and digital dividends.
3. ICTs for agriculture and rural development.
4. Concept of IT convergence in agriculture extension: Cyber extension.
5. Harnessing ICT for Agriculture and Rural development: Indian initiatives.

Genesis of Personal Computers and Internet

It is relevant in this context to trace the history of personal computers and Internet briefly. In this regard, efforts have been made to present this content in simple language.

Computers

Computer is a machine that performs calculations and process information with astonishing speed and precision. A computer can handle vast amounts of information and solve complicated problems. The physical equipment that makes up a computer system is called hardware, which includes input and output devices, file storage devices, the memory and the processor. Computer software consists of the programs that a computer uses to perform a task.

In 1888, American businessman Herman Hollerith devised a punched card system for tabulating the results of the United States census. In 1930, American electrical engineer Vannevar Bush devised a differential analyzer to solve differential quotations. The first special purpose electronic digital computer was constructed in 1939 by John V. Atanasoff, an American mathematician and physicist. In 1944, Howard Aiken, a Harvard University professor, built an early form of computer called Mark I. In 1951, Eckert and Mauchly invented a more advanced computer called Universal Automatic Computer (UNIVAC Internet). In 1958, Control Data Corporation introduced the first fully transistorized computer, designed by American engineer Seymour Cray. The first personal computer was introduced in 1975 and in 1977, two American students Steven P. Jobs and Stephen G. Woznaik introduced the Apple II personal computer, which was much less expensive than mainframes. As a result, computers became available to people other than computer specialists and technicians.

The Internet

Internet is a self-regulated network of computer networks connecting millions of computer systems all over the world, where one can communicate text, audio, video and animation (Turban *et al.*, 2000).

The Internet was conceived in 1963 by Larry Roberts, while he was working for the Advanced Research Project Agency (ARPA) with funding from the United States Department of Defence (Negroponte, 1995a and 1995b). ARPAnet, as it was called in the beginning, emerged as a communication tool for a handful of defence workers and contractors. It was designed as a fundamentally decentralized network in order to be a fail-safe communication system (Richardson, 1995). The wife of one of the ARPAnet engineers was hearing impaired and he wanted to communicate from office with her at home, for which he used a telephone text messaging device enabled him and his wife to converse through text messages sent through the telephone system. This has influenced the engineer to incorporate text messaging between people as a feature of ARPAnet, this feature is now popularly known as Electronic mail (e-mail).

In 1970s, students, faculty members of universities and scientists from research laboratories began to use this popular tool. In 1980s, National Science Foundation of United States developed a network called NSFnet, through which many universities and colleges were connected. The more recent tool and the one that has catapulted the Internet into popular culture, is the World Wide Web. The WWW is the part of Internet where one can witness the emergence of a vast global library of information.

Digital Divides and Digital Dividends

Some enthusiastically claim that the IT has the capacity to bring knowledge and hence the prosperity is isolated to marginalized individuals and nations. Others fear that those individuals and nations that are already most privileged, will be ones who will benefit from the IT, leaving others to fall even further. No where is the debate between optimists and pessimists more relevant than India, where new centres of IT are creating wealth and integrating some sectors into global economy, while millions remain marginalized and deprived of the most basic human needs. No developing country has benefited more from the digital revolution than India, and in no country is the digital divide wider and deeper.

Digital Divides at Macro Level

In practice, two biggest hurdles to achieve uniformity in gains from the IT industry are access and content. Access to IT is in turn obstacled by linguistic and technological barriers. Linguistic barrier limits the access to that medium as terms of understanding and utilizing the medium. A divide between English and non-English speaking communities, where comprehensible content is available only to a selected few who have knowledge of English. As it is seen in Indian scenario, most of the users of IT belong to elite class who are fluent in English and other sections of Indian society have been facing linguistic barrier, as they are not in a position to comprehend the contents available in English and technological barriers to access also influence control over usage of IT (Kenesten, 2001).

Consequently, the barriers to access have an obvious influence on the type of content relayed through IT. Since the deprived and non-elite sections do not have access to the IT services, the web sites provide only such type of content which is targeted at those who have access to the IT services.

The circle in Fig, 5.1 has to be broken in order to improve the access and content parts of IT with respect to the grass roots level people. Because of the complex relationship between barriers of access and content without more and easier access, content will continue to be fewer representatives, which leads to less incentive for under presented groups to overcome the access barrier. This is perhaps the most important aspect of digital divide.

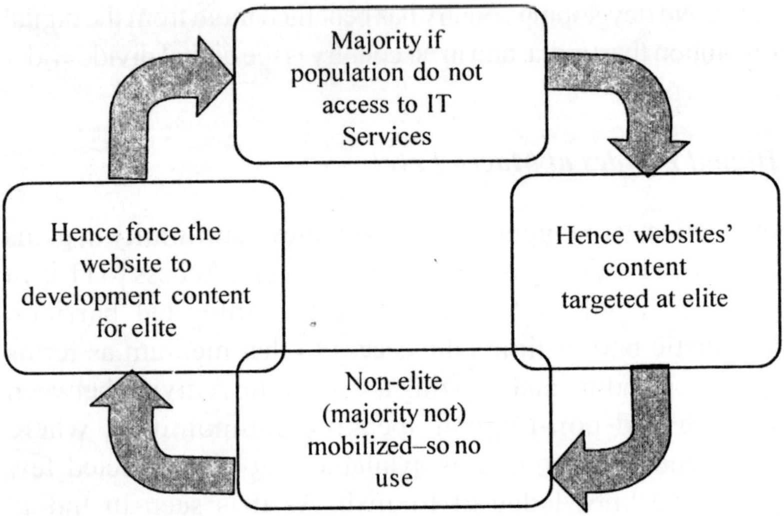

Fig. 5.1: Viscius circle of access and content complex relationship

North-South Divide

The north-south digital divide flows from growing digital gap between the rich and poor nations. At one extreme end there are the United States and the Nordic countries like Germany, Sweden, Finland and Iceland, in 2000 had over 90 per cent telephone connectivity, home based computer ownership is over 50 per cent and Internet connectivity is averaged at least 40 per cent. At the other extreme lies Africa, South America and South Asia, where telephone connectivity is 30 per cent, home based computer ownership is 1-2 per cent and Internet connectivity very meagre (Arunabha and Nandan, 2001). This led to widening of already great gulf existing between North and South.

Infrastructure Barriers and Building Bridges: Case of India

The Infrastructure bottlenecks in case of India which hinder the spread of IT to villages have to be considered in order to conceptualize the digital divide. The Infrastructure bottlenecks can be divided into three aspects, viz. connectivity, electricity and lack of local language software.

Telephone density is nudging just 3 fixed connections per 100 people, of which more than a fifth is in metros (Down to Earth, 2001). The Indian Institute of Technology, Chennai created high speed and low cost Internet access system, which allows for dense high speed connectivity across long distances, that allow telephone and Internet access to end users.

Another most important constraint in Indian context is poor electricity supply. Acute power cuts in rural areas, whereas power generation and transmission record is very poor; it is, therefore, an uphill task to meet the demand of IT in villages. The Indian Institute of Science, Bangalore designed a hand held Simputers, which will run on battery and provide Internet and e-mail access.

Third obstacle is the lack of local language software. Computer and IT would be completely useless in villages if their content is in English. A country which has so many local languages, the large scale development of local language software gains importance. The Centre for the Development of Advanced Computing (CDAC) has recently developed high quality Indian language software that affords for text processing, web designing and translation system between English and other languages and among Indian languages (Sood, 2000).

Digital Divide: Farmers Perspectives

The IT is going to make a change in the rural areas in general and the farming communities in particular. Slow and steady IT interventions (both private and non-governmental) are going to occur to influence the information base of the farming communities. If the trend continues for a decade or so, the commercial and large farmers who have hitherto been exposed to different information sources, will have the easy access to IT services and small and marginal farmers particularly in complex, diverse, risk prone (CDR) regions will again be deprived of the needed information. This trend will result in widening gap between large and small, marginal farmers and a divide will also be seen between the well endowed regions and the CDR regions.

ICTs for Agriculture and Rural Development

With regard to ICTs (in particular Internet) use in support of agriculture and rural development, applications fall into five main areas: economic development for agricultural producers, community development, research/education, small and marginal enterprise (SME) development and media networks (FAO, 1996).

Economic Development Applications for Agricultural Producers

Rural communities and small-scale agricultural producers are deeply affected by global economic, environmental and political forces. The idea that communities of small-scale agricultural producers are isolated and living in closed, self-sufficient societies is a myth. In fact, with the appropriate knowledge, small-scale producers can even have a competitive edge over larger operations. When knowledge is harnessed by strong organizations of small producers, strategic planning can be used to provide members with lower-cost inputs, better storage facilities, improved transportation links and collective negotiations with buyers. The Internet is one tool that can enhance this flow of information. It is an inexpensive way to communicate and access global information. Local Internet services can be easily managed by well-organized local user groups and farmers' organizations. Information and analyses can be tailored to local, regional and national knowledge and communication needs and realities. When combined with national and global market information systems, and with the ability to communicate quickly with potential buyers and brokers, local Internet systems become valuable strategic planning and decision-making tools.

Community information centres and farmers' organizations can also gather information from the Internet and disseminate it via local radio stations, newspapers and other local information-sharing networks and tools. For example, daily market prices and agricultural news can be posted at cooperatives, local stores, transportation hubs, agricultural supply outlets and social gathering points. Simple newsletters can be developed using Internet information and distributed to members of farmers' organizations. When integrated with other media tools, the Internet can be a powerful information resource and research tool.

Community Development Applications

Internet services are also valuable when placed at the service of rural development-oriented organizations which act as local communication conduits or intermediaries. Along with providing improved market knowledge, they can also develop locally appropriate applications and creative services; provide knowledge about successful development strategies, enable efficient regional, national and global organizational efforts (the use of the Internet as a global communication and organizing tool in Chiapas, Mexico, is an excellent example of the latter); improve access to a huge variety of information, training, research and educational resources (including distance education services) which are typically unavailable in rural and remote areas because of the costs associated with printed materials and books; enable rural young people to learn about computers and to have access to the technologies and information available to their urban peers; provide access to critical technical information for rural professionals such as physicians, health care workers, technicians and engineers, thereby providing further encouragement for these professionals to continue practicing in rural and remote communities; used as marketing tools to promote rural tourism and market the products of small secondary industries and home-based businesses; enable local NGOs to gain a global presence and make better contact with potential donors and supporters through the on-line publication of resources and information and through the use of electronic mail and sensitize urban policy-makers to the realities and needs of rural populations.

Research/Education Applications

Within national, regional and international research communities, increased attention has been directed towards "participatory research" strategies (Chambers and Guijt, 1996 and FAO, 1995). These strategies place farmers and rural residents at the centre of the research process and enable them to enrich their knowledge base and share it with one another as well as with fieldworkers, researchers and decision-makers. Internet use among intermediary organizations and leaders involved in participatory research can provide a cost-effective method for documenting and sharing lessons learned and research results. Internet use also has the potential to strengthen linkages between and among farmers' organizations, extension workers, researchers and policy-makers.

The cost of accessing printed academic materials within developing countries is usually so high that students and faculty members have great difficulty in acquiring books and journals. Furthermore, the time required to obtain printed materials from overseas can be long enough to render some information outdated by the time it arrives. Via the Internet, any information published on-line can be accessed almost instantly and at a fraction of the cost of obtaining printed materials. Archival lists of resources can be easily reviewed and assessed in remote locations.

Electronic distance education services are already in use in North America, Australia and Europe (particularly among people in rural areas), and with the continued growth of Internet access in developing countries there is a very good chance that similar services will develop a significant demand. On the African continent, Internet-based distance education programs are already underway in Ghana, South Africa and Egypt. Distance education (as well as traditional education) partnerships between universities in the North and the South (such as the University of Guelph's partnerships with universities in Cameroon and India to develop distance education extension worker training programs) have proved to be beneficial to the institutions involved. With the assistance of Internet tools, these partnerships can be further strengthened and Internet learning resources can be cooperatively developed across oceans to be utilized by participants in developing nations. Of course, this process can work in the other direction too, to enable students in the North to learn more about the conditions, challenges, potentials and knowledge development of the South.

Overall, the Internet holds significant potential to enhance learning and research relationships among researchers, academics and students wherever they may be located. The list of potential applications is infinite and thousands of informal linkages of this sort take place everyday in Internet discussion groups. Development agencies such as FAO can play a role in helping to formalize and provide credentials and diplomas for people who participate in specific electronic learning initiatives delivered via the Internet.

Small and Medium Enterprise Development

The removal of international trade barriers has brought quickly changing global markets. Large international corporations can now compete for the SMEs' market, but SMEs traditionally have not had the infrastructure or necessary resources to fight back. Our mission is to provide a productive and professional Internet/WWW-based network to help SMEs communicate about business needs, share their resources and expand their markets. Private sector businesses, large and small, are using the Internet to reach new markets, promote products and services globally and access critical business and financial information.

The tourism sector has been quick to recognize the benefits of the Internet for advertising destinations, tours and holiday services. Of particular interest are the World Wide Web sites for "ecotourism", game parks and adventure tours in areas of southern Africa where rural tourism is a growing industry. Tourism operators in rural and remote areas have a difficult time marketing their destinations through traditional media owing to production and distribution costs. The Internet now represents a very inexpensive way for them to show their sites to the world and interact directly with potential tourists.

News Media Networks

The news media in developing countries have also been at the forefront of developing Internet applications. For example, in Zambia, national daily newspapers mirror their daily copies on the World Wide Web, making the local news accessible to expatriate Zambians around the world. E-mail discussion groups provide these expatriates with an opportunity to discuss the daily news with one another and with their Internet-connected peers in Zambia. A discussion group joined by the author generated a minimum of 30 e-mail messages per day Such e-mail discussion groups for expatriates and nationals exist for virtually every developing country in the world and represent a relatively untapped resource for accessing the views, ideas and creativity of members of civil society with regard to development policy and initiatives.

In addition to the latter news and information applications, organizations such as Inter Press Service (IPS) Third World News Agency use the Internet to source news stories from local writers in developing countries and share those stories with international wire services such as Associated Press. IPS is also able to provide Internet feeds that enable African news media to have access to African news from around the continent. This is particularly relevant to rural radio stations and other rural newspaper and newsletter producers that would otherwise be unable to obtain the same news from other sources. IPS can also provide an outlet for rural news writers to share their stories regionally, nationally and globally. Similar Internet strategies for rural radio networks, which might also incorporate digital audio transmissions, may well emerge in the near future.

Concept of IT Convergence in Agriculture Extension: Cyber Extension

The new agricultural paradigm in India will have to be recast to take advantage of knowledge availability to achieve the triple goals of increased income, jobs and food security. The emerging ITs have a significant role to play in evolving such paradigm, as was evident from an interdisciplinary dialogue on "Information Technology: Reaching the Unreached" held in 1992.

Agricultural extension is now becoming more diversified, more technology intensive and more demand driven. With many potential applications, IT can help the extension system to be more effective in meeting farmers information needs (Zijp, 1994). Information Technology will bring new information services to rural areas and farmers, as users will have much greater control than ever over current information channels, i.e. agricultural extension system. This will lead to realization of much talked about "Cyber Extension", which includes effective use of ICTs, Internet, Intranet, Expert systems, Multimedia learning and Computer based training systems to improve information access to the farmers, extension workers, research scientists etc.

Some of the potential advantages of cyber extension over traditional extension are worth mentioning in this context and they are, viz. cyber extension will save money, time and efforts; cut steps from extension process; be information rich and interactive; offer instant international reach and be continuously available (Sharma, 1999).

Harnessing ICT for Agriculture and Rural Development: Indian Initiatives

For better understanding, this section of the chapter is presented in the following three sub-heads:

Success Stories

There are cases of application of ICTs, that have made a difference in the delivery of services in rural India. Some of these include:

Warna Wired Village Project

Warna Nagar is cluster of 70 villages consisting 46 villages from Kolhapur and 24 villages from Sangli districts of Maharashtra is an eye of the wired villages project. In 1960, Shri Tahasaheb Kore propagated the idea of cooperatives as a method of achieving socio-economic development and thus born the Warna Nagar Cooperative Society. In 1998, Mr. Vinay Kore initiated the project, jointly implemented by Government of India, Government of Maharastra and Warna Cooperative Society with the share of financial support being in the ratio of 50:40:10. Currently 46 computer booths are functioning in Kolhapur and meeting the information needs of the farmers, remaining 24 villages of Sangli district are waiting to link to central server station. The project aims to:

- Utilize IT to increase the efficiency/productivity of the existing cooperative society.
- Provide greater transparency in the working of cooperative society.
- Provide agiculture, animal husbandry, medical and educational information to villagers.
- Bring knowledge of the world to the doorsteps of Warna.

Information Villages Project

The MS Swaminathan Research Foundation is aimed at bringing the benefits of modern information and communication technologies to rural families in Pondicherry. A Value Addition Centre, which is the hub of the information network, has been established in Villianur village and four information shops have been established in different villages.

National Dairy Development Board

IT-based machines are being used at milk collection centres, and in cooperatives to measure butter/fat content of milk, test the quality of milk, and promptly make payment to the farmers. It has resulted in the removal of incentives to cut the milk by adding water, reduced time for payments from 10 days to less than five minutes, and instilled confidence in farmers in the cooperative set up. All of these factors have helped the milk market to expand.

Honey-Bee Knowledge Network

Information and Communication Technologies (ICT) can help empower the knowledge rich but economically poor people. Indian Institute Management, Ahmedabad used to augment grassroots inventors and overcome language, literacy and localism a large number of grassroots inventions have been identified and documented as short multimedia presentations. Future plans include creating a database of such innovations and making them accessible via a wide area network.

Indian Space Research Organization

One way video, two way audio teleconferencing interactive networks have been used for education and training by this scientific organization. The major application of the network in rural development was for training extension staff from various departments of the State Governments. In addition, a large number of women, Panchayati Raj elected officials, primary school teachers, and child development workers spread over large distances have been trained.

Gyandoot Project

In 2000 at Dhar district (More than 60 per cent of the inhabitants live below poverty line, around 54 per cent belongs to scheduled tribes) of Madhya Pradesh, the project has installed a low-cost, self-sustainable and community-owned rural Internet centres that caters to the specific needs of village communities. There were initial difficulties in acceptance of this project. However, the project has been sustained by the demands of the users and strong support of the local bureaucrats.

Babhaleswar Project

In 1999 at Babhaleswar (Ahmednagar district of Maharashtra) the high speed network connectivity established at Krishi Vigyan Kendra with an objective to provide information services like technical crop cultivation practices, disease control, market information, weather forecast, health, e-mail, schemes and programmes of State and Central Governments etc. Apart from KVK's Internet access, network also has ARIS cell set-up under ICAR's NATP project for the access of information. This has been successful in increasing the involvement of not only the farmers but also the farmer women and rural youth.

Andhra Pradesh State-wide Area Network (APSWAN)

Aims to link the state government's secretariat with 23 district headquarters, serving as the backbone for "multimedia-services" (voice, video and data) that would be used for improved co-ordination between state headquarters and district offices in managing various regulatory, developmental, and hazard mitigation programmes of the State Government. Mandals will be served by this two-way communication, and electronic commerce applications will be developed. The Andhra Pradesh Value Added Network Services project hopes to deliver a variety of public services through a large network of information kiosks.

Information and communication technologies are an important part of the Government of Andhra Pradesh's efforts to improve the efficiency of its administrative offices. AP is the first state in India to design a statewide computerization program that will be used in rural areas, at the mandal-level (the administrative unit above the village-level panchayat). There are 1124 mandals in the state. The first software application is the issuance of certificates pertaining to land holdings, caste, nativity and income across a common counter, without the current delay of 15 to 20-days. The Computer-aided Administration of Registration Department (CARD), a project of Andhra Pradesh aims to introduce a transparent system of property valuation, which is easily accessible to citizens.

Understanding the Clientele and their Information Needs

In IT initiatives aimed at agricultural development, the stakeholders are farmers and it is common to find that intended users (farmers) are unaware of the ICT services being developed for them. Developing information services with farmers and catalyzing sustainable user management of communication networks is not a common strategy among funding agencies. Current evidences suggest that to achieve sustainability and success, Internet projects for rural and agricultural development must began with the real needs of the local farmers (Richardson, 1996).

Socio-economic factors, psychological factors and agro-climatic aspects of farmers are first and foremost aspects, which are prerequisites of understanding farmers. The appropriateness of the prospective services is to be rated with respect to the farmers' needs.

Understanding Impact Process of ICT Projects

In the Indian context, perhaps for last few years attempts have been made to publish the ICT projects working towards agriculture and rural development. Most of the attempts simply described the details of functioning of ICT projects without much emphasis on the impacts that projects make on the community. This is coupled with the fact that the evaluation of ICT projects on communities is still an open issue and it is evident from the proceedings of workshop on "Telecentre Evaluation: A global perspective" (1999), universally accepted norms/methods for quantification of impacts of ICTs are yet to be defined.

In this background, it is essential to focus research on impacts not only on ICT users and local applications, but also on people who do not participate directly in local ICT initiatives and enable them to benefit directly or indirectly from these initiatives (Richardson, 1996).

RESULTS AND DISCUSSION 6

Keeping in view the objectives and the empirical evidence obtained through research, the results have been analyzed by using appropriate statistical and analytical tools. The findings arrived at, have been presented and discussed, considering the theoretical and available research in the area of investigation. While presenting the results, the related aspects have been clubbed into different sections and finally, at the end of each section discussion of these results have been attempted.

1. Personal and livelihood characteristics of livestock farmers.
2. Information seeking behaviour of livestock farmers.
3. Internet characteristics of livestock farmers.
4. Kind of information available through Internet, its utility and credibility.
5. Knowledge and perceived benefits of livestock farmers attained through Internet.
6. Relationship between dependent and independent variables.
7. Problems faced and suggestions given by the livestock farmers to improve effectiveness of Internet communication.

Personal and Livelihood Characteristics of Liverstock Farmers

Personal Characteristics of Livestock Farmers

The personal characteristics of livestock farmers include age, gender and socio-economic status.

Age: The age of the respondents ranged between 24 to 59 years and the overall average age, was found to be 41.08 years. Majority (57.86%) of the Internet users belonged to middle age, while 23.57 per cent Internet users were from younger age group and only 18.57 per cent belonged to old age group (Table 6.1 and Fig. 6.1).

Table 6.1: Distribution of the respondents according to their age (in years)

S.No.	*Category*	*Frequency*	*Percentage*
i.	Young (24-33)	66	23.57
ii.	Middle (34-48)	162	57.86
iii.	Old (49-59)	52	18.57

Mean: 41.08 S.D.: 7.91

By this, one could conclude that middle aged farmers were considered to be matured and are more positive in using the Internet services, followed by young and old age groups. There is relatively little information to predict the effect of reported age on technology interventions. In fact the oldest and youngest groups arrayed similarly. Possibly this is because many younger livestock owners enter the automation milieu with the same initial tendencies of anxiety and fear that are often attributed to older farmers. These results were in line with the observations of Ruth (2000).

Gender: It was evident from Table 6.3, that majority (77.50%) of Internet users were male and the female comprises 22.50 per cent of the total respondents (Fig. 6.1).

Table 6.2: Distribution of respondents as per the gender

S.No.	*Category*	*Frequency*	*Percentage*
i.	Male	217	77.50
ii.	Female	63	22.50

One frequently noted dimension of inequality in Internet access and usage is gender. Many studies, viz. Shashaani (1997) and Sherman *et al*. (1998) carried out on Internet usage noted that women were less likely than men to use the Internet and men tend to be more intended in computer use than women. These findings were in conformity with the studies of Edwards and Wajcman (1995) and Shashaani (1997).

Socio-economic Status: The socio-economic status scale in turn included thirteen items as shown in Table 6.3.

Table 6.3: Distribution of respondents according to their socio-economic status

S.No.	*Category*	*Frequency*	*Percentage*	*Mean (S.D.)*
I.	**Religion**			
i.	Hindu	204	72.90	
ii.	Muslim	48	17.10	
iii.	Christian	28	10.00	
II.	**Caste**			
i.	SC/ST	54	19.29	
ii.	BC	76	27.14	
iii.	OBC	71	25.36	
iv.	General	79	28.21	

Contd.......

S.No.	*Category*	*Frequency*	*Percentage*	*Mean*	*(S.D.)*
III.	**Self-education**				
i.	Illiterate	0	0.00		
ii.	Functionally literate	9	3.22		
iii.	Primary School	12	4.28		
iv.	Middle School	38	13.57		
v.	High School	75	26.79		
vi.	Intermediate	78	27.85		
vii.	Graduate and above	68	24.29		
IV.	**Family educational status**				
i.	Low (<2)	47	16.79	3.10	(1.03)
ii.	Medium (2 to 4)	154	55.00		
iii.	High (>4)	79	28.21		
V.	**Marital status**				
i.	Unmarried	38	13.57		
ii.	Married	242	86.43		
VI.	**Family type**				
i.	Nuclear	133	47.50		
ii.	Joint	147	52.50		
VII.	**Family size**				
i.	Small (1-3)	42	15.00		
ii.	Medium (4-6)	68	24.28		
iii.	Large (7-9)	95	33.93		
iv.	Very large (>9)	75	26.79		
VIII.	**Material possession**				
i.	Low (<8)	48	17.14	9.58	(1.78)
ii.	Medium (8 to 11)	204	72.86		
iii.	High (>11)	28	10.00		
IX.	**Communication aids**				
i.	Low (<8)	31	11.07	11.50	(3.50)
ii.	Medium (8 to 14)	167	59.64		
iii.	High (>14)	82	29.29		
X.	Type of house				
i.	Thatched shed	0	0.00		
ii.	Mudwalls & thatched	59	21.10		
iii.	Brickwalls & tiled	159	56.80		
iv.	Concrete house	55	19.60		
v.	Concrete double storied	7	2.50		

Contd.......

S.No.	*Category*		*Frequency*	*Percentage*	*Mean (S.D.)*
XI.	**Position in the family**				
i.	Head		109	38.93	
ii.	Independent		153	54.64	
iii.	Dependent		18	6.43	
XII.	**Self-occupation**				
i.	Livestock farming		252	90.00	
ii.	Agriculture		19	6.79	
iii.	Agricultural labourer		0	0.00	
iv.	Service		9	3.21	
XIII.	**Family members' occupation**				
i.	Livestock farming		173	61.79	
ii.	Agriculture		37	13.21	
iii.	Agriculture laborer		40	14.28	
iv.	Service		30	10.72	
	Overall				
i.	Low	(<25)	46	16.43	29.85 (4.95)
ii.	Medium	(25 to 35)	182	65.00	
iii.	High	(>35)	52	18.57	

Religion: Table 6.3 indicated that 72.90 per cent of the respondents were Hindus, while 17.10 per cent of the respondents belonged to Muslim category and 10.00 per cent were Christians.

These results infered that wide variety of acceptance of new technologies were not by the members of specific ethnic groups but the population as a whole (i.e. Hindu group). The above results were in similar lines to observations of Nie *et al.* (2001).

Caste: Table 6.3 indicated that more or less number of Internet users were from general castes (28.21%) and backward castes (27.14%) whereas relatively less number of other backward castes (25.36%) category and only 19.29 per cent of Internet users were from scheduled caste/scheduled tribe category. This is, because, livestock farming is no more a caste occupation and has proved to be the most beneficial for all categories of farmers. The above findings were in contrast with the observations of Charan and Sharma (1989) and Manjunatha (1991), their studies might have involved lower strata or caste groups.

Self-education: It was apparent from Table 6.3, that there are no illiterates among the respondents. Majority (27.85%) of the respondents were educated up to Intermediate level, followed by High School (26.79%), Graduate and above (24.29%), Middle School (13.57%), Primary School (4.28%) and 3.21 per cent of respondents were functionally literate.

The above results implied that the Internet access and usage demands the literacy among the users. Thus, none of the respondents belonged to illiterate group and an overview indicates that general literacy level was high among the livestock farmers using Internet services. These results were in line with the findings of Khaldoun Almhanna (1998).

Family educational status: The data presented in Table 6.3 pointed out that majority (55.00%) of the respondents were having medium family educational status, followed by high (28.21%) category and low (16.79%) family educational status category.

The Internet access as well as usage demands the literacy among the users. An overview of above results indicated that the literacy level among the family members were very fair, this may be the reason for the family members' unhesitant approach to Internet.

Marital status: On perusal of Table 6.3, it was observed that majority (86.43%) of the respondents belonged to married category followed by 13.57 per cent who were unmarried.

Family type: It was evident from Table 6.3, that majority (52.50%) of the respondents were from joint family, while 47.50 per cent of respondents were from nuclear family.

These results imply that livestock farming is labour intensive and suits to the large sized rural families. These findings were in contrast to the reports of Ramchand *et al*., (1990) and Rao (1982).

Family size: It was apparent from Table 6.3, that majority (33.93%) of the respondents possessed large, followed by very large (26.79%) and medium (24.28%) families and only 15.00 per cent of the respondents had small family size.

These results imply that livestock farming is labour intensive and effective family members involved in this occupation enlarge the family size in rural areas. The above results were in similar lines of the observations of Manjunath and Marilingannavar (1989).

Material possession: It was observed from Table 6.3, that majority (72.86%) of the respondents had medium material possession; while 17.14 per cent had low material possession and 10.00 per cent of the respondents had high material possession.

Communication aids: It was observed from Table 6.3, that majority (59.64%) of the respondents possessed medium communication aids, while 29.29 per cent had high communication aids and 11.07 per cent of the respondents possessed low communication aids.

The selected Internet users in this study possessed different varieties of communication aids, viz. radio, television, tape recorder, V.C.R., V.C.D., telephone, computer etc.

Type of House: It was clear from Table 6.3, that majority (56.80%) had brick walls with tiled type of house, followed by mud walls with thatched roof (21.10%), concrete (19.60%) and 2.50 per cent of the respondents had concrete double storied type of house. No livestock farmer had thatched shed type of house.

Position in the family: It was observed from Table 6.3, that majority (54.64%) of the respondents were independent; while 38.93 per cent were heads and only 6.43 per cent were dependent. This is because; most of the respondents belonged to middle and younger age group and were independent in nature.

A good number of respondents belonged to nuclear family and this might be the cause for more number of independent individuals. Similar observations were made by Samantha (1993).

Self occupation: It was observed from Table 6.3, that most (90.00%) of the respondents were found to have livestock farming as their occupation followed by agriculture (6.79%) and only 3.21 per cent of the respondents were serving in various government or non-government organizations. None of the respondents were agricultural labourers.

The main reason that has motivated the respondents to go for livestock farming was its economic and employment potential. There is a proportionally higher rate of decline in the population engaged in agriculture compared to the rate of decline in percent of rural population. However, the percentage of population engaged in livestock farming have been increasing steadily (Agricultural Statistics, 2001).

Family members' occupation: It was apparent from the Table 6.3, that majority (61.79%) of the respondents were found to have livestock farming as their family members' occupation, followed by agriculture labour (14.28%), agriculture (13.21%) and service (10.72%).

The family members of the respondents had gone for variety of occupations apart from livestock farming and fair percentage of family members undertook agriculture, labour and even services as their occupation. This could be due to their higher outreach and ambitions in the field of livestock production.

The overall socio-economic status of the respondents was presented in Table 6.3, which revealed that majority (65.00%) of the respondents had medium socio-economic status followed by high (18.57%) and low (16.43%) categories (Fig.6.3).

From the above results it could be inferred that the livestock farming and IT are such enterprises that can be effectively undertaken by the farmers of average socio-economic status. These findings were in agreement with the observations of Murthy (1990) and Ramchand *et al*., (1990).

Livelihood Characteristics of Livestock Farmers

Land holding: It was clear from Table 6.4, that majority (56.78%) of the livestock farmers were landless. 15.36 per cent of the livestock farmers belonged to marginal, followed by small (12.50%), low medium (9.30%), large (5.36%) and 0.70 per cent belonged to high medium category.

Table 6.4: Distribution of respondents according to their land holding

S.No.	*Category*		*Frequency*	*Percentage*
i.	Landless		159	56.78
ii.	Marginal	(< 1ha.)	43	15.36
iii.	Small	(1.1 to 2 ha)	35	12.50
iv.	Low medium	(2.1 to 4 ha.)	26	9.30
v.	High medium	(4.1 to 10 ha.)	2	0.70
vi.	Large	(> 10 ha.)	14	5.36

The above results are, due to the fact that the respondents were livestock farmers with livestock farming as main occupation and about 56.78 per cent of the respondents were livestock farmers without any land. Investigator learnt that livestock farmers were using the land for fodder cultivation to feed their animals economically and it was also evident that the livestock farming was seen as an enterprise rather than a means of livelihood of landless rural poor.

Herd Size: It was apparent from Table 6.5, that among the cattle owners majority (59.50%) had medium herd size, while 26.45 per cent of the cattle owners had small herd size and 14.05

per cent possessed large herd size. Among buffalo owners, majority (68.49%) had small herd size, while 23.29 per cent of the buffalo owners had medium herd size and 8.22 per cent possessed large herd size.

Table 6.5: Distribution of respondents according to their herd size (species-wise)

S.No.	*Category*	*Frequency*	*Percentage*	*Mean (SD)*
A.	**Cattle (n = 121)**			
i.	Small (<5)	32	26.45	10.34 (5.57)
ii.	Medium (5 to 16)	72	59.50	
iii.	Large (>16)	17	14.05	
B.	**Buffalo (n = 73)**			
i.	Small (<4)	19	26.03	5.19 (1.20)
ii.	Medium (4 to 6)	37	50.68	
iii.	Large (>6)	17	23.29	
C.	**Sheep (n = 40)**			
i.	Small (< 150)	11	27.5	275.50 (125.53)
ii.	Medium (150 to 400)	23	57.5	
iii.	Large (>400)	6	15.00	
D.	**Poultry (n = 46)**			
i.	Small (<3000)	8	17.39	5956.52 (2897.94)
ii.	Medium (3000 to 9000)	25	54.35	
iii.	Large (>9000)	13	28.26	

It was also apparent from Table 6.5, that among sheep owners, majority (57.50%) possessed medium flock size, while 27.50 per cent of the sheep owners had small flock size and 15.00 per cent possessed large flock size. Whereas, in case of poultry owners, majority (54.35%) had medium flock size, followed by large (28.26%) and small (17.39%) flock size (Fig. 6.8).

Rearing systems: It was observed from Table 6.5, that majority (42.98%) of cattle owners followed Intensive type of rearing pattern, while 31.40 per cent followed range pattern and 25.62 per cent of cattle owners followed semi-intensive rearing pattern. Among buffalo owners, majority (45.21%) of buffalo owners followed semi-intensive type of rearing pattern, while 42.46 per cent followed range pattern and 12.33 per cent of cattle owners followed intensive rearing pattern. Among sheep owners, majority (72.50%) of sheep owners followed range type of rearing pattern, while 27.50 per cent followed semi-intensive pattern and none of the sheep owners followed intensive rearing pattern. Whereas, among poultry owners majority (73.91%) followed deep litter system rearing and 26.09 per cent of the poultry owners followed cage system. The differences in rearing pattern may be attributed to variation in species, size of the herd/flock and animal feeding habits.

Table 6.6: Distribution of respondents according to rearing systems

S.No.	*Category*	*Frequency*	*Percentage*
A.	**Cattle (n = 121)**		
i.	Range	38	31.40
ii.	Semi-intensive	31	25.62
iii.	Intensive	52	42.98
B.	**Buffalo (n = 73)**		
i.	Range	31	42.46
ii.	Semi-intensive	33	45.21
iii.	Intensive	9	12.33
C.	**Sheep (n = 40)**		
i.	Range	29	72.50
ii.	Semi-intensive	11	27.50
iii.	Intensive	0	0.00
D.	**Poultry (n = 46)**		
i.	Deep litter	34	73.91
ii.	Cage	12	26.09

Feeding patterns: It was evident from Table 6.7, that majority (68.40%) of the cattle owners fed their herd both roughages and concentrates, while 31.40 per cent cattle owners feed only roughages to their herd. Among buffalo owners, majority (57.54%) fed roughages and concentrates and 42.46 per cent feed only roughages to their herd. Among sheep owners, majority (72.50%) of the sheep owners gone for only grazing system, followed by and 27.50 per cent of the sheep farmers followed the feeding pattern of grazing along with supplementation. Whereas, among poultry owners majority (63.04%) fed starters and growers to their flock and 36.96 per cent of poultry owners fed starters, growers and finishers.

Table 6.7. Distribution of respondents according to feeding patterns

S.No.	*Category*	*Frequency*	*Percentage*
A.	**Cattle (n = 121)**		
i.	Only roughages	38	31.40
ii.	Roughages & concentrates	83	68.40
B.	**Buffalo (n = 73)**		
i.	Only roughages	31	42.46
ii.	Roughages & concentrates	42	57.54
C.	**Sheep (n = 40)**		
i.	Grazing and supplementation	11	27.50
ii.	Only grazing	29	72.50
D.	**Poultry (n = 46)**		
i.	Starter & grower	17	36.96
ii.	Starter, grower & finisher	29	63.04

The differences in feeding pattern may be attributed to variations in traditional animal feeding habits, size of the herd/ flock, production level and purpose of livestock rearing.

Total Income: It was clear from Table 6.8 that among livestock farming families, majority (46.54%) cattle owners followed by buffalo owners (25.34%), poultry (16.60%) and only 11.52 per cent of the total were sheep farmers. This could be due to their more involvement in cattle and buffalo farming in comparison to poultry and sheep farming.

In aggregate, majority (77.50%) of the respondents had earned their total income through livestock farming, while 8.22 per cent of the respondents had earned their total income through agriculture. Only 7.14 per cent of the respondents had earned through laborer as well as service (Fig. 6.1).

Table 6.8: Distribution of respondents according to species-wise contribution to total income

S.No.	*Category*	*Below 1 lakh*	*1.0 to 3.0 lakhs*	*Above 3 lakhs*	*Total*	*Percen-tage*
A.	**Livestock**	74	115	28	217	77.50
i.	Cattle	40	52	9	101	46.54
ii.	Buffalo	18	32	5	55	25.34
iii.	Sheep	16	9	0	25	11.52
iv.	Poultry	0	22	14	36	16.60
B.	**Agriculture**	0	14	9	23	8.22
C.	**Labourer**	18	2	0	20	7.14
D.	**Servicentage**	16	4	0	20	7.14
	Total	108	135	37	280	100.00

The differences in income gain could be due to the variation in herd size, farm size, knowledge, adoption and extent of involvement of family members in various income generating activities.

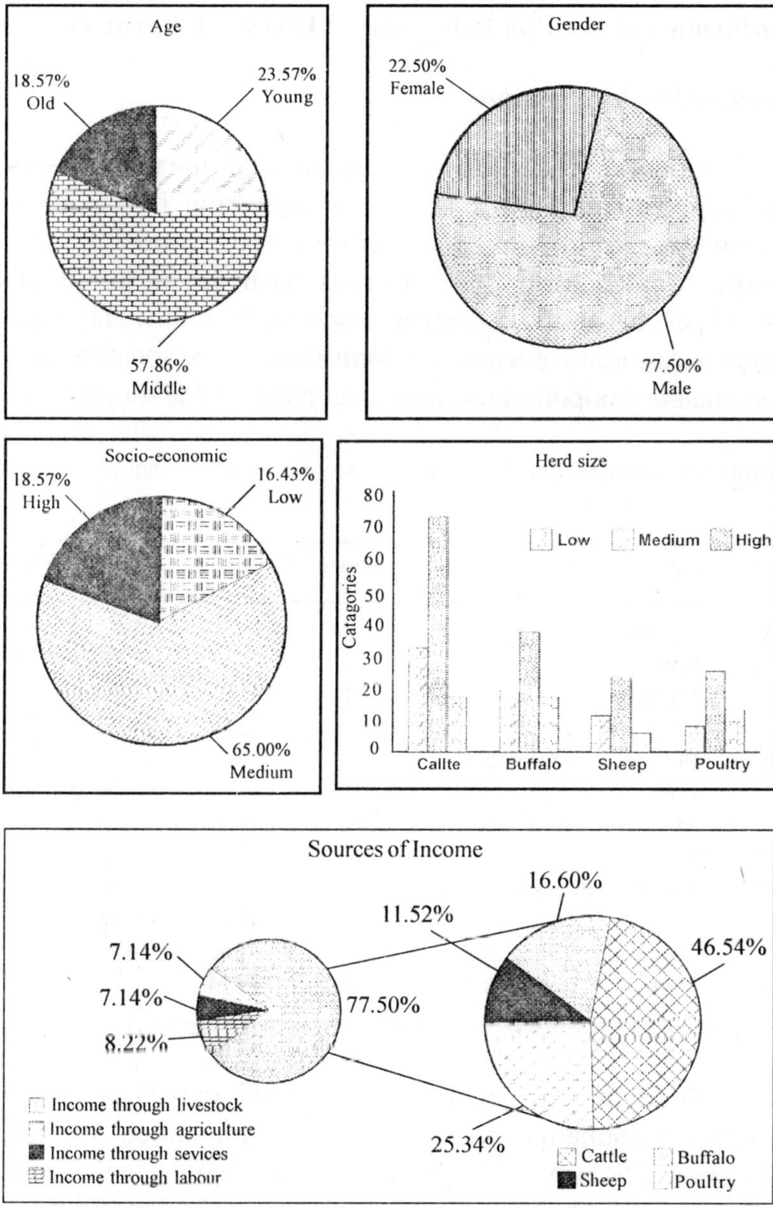

Fig. 6.1: Distribution of respondents according to their personal characteristics

Information Seeking Behaviour of Livestock Farmers

Sources of Information

It is evident from Table 6.9, that majority of the respondents belonged to medium category in acquiring information from personal-localite, personal-cosmopolite and impersonal-cosmopolite sources with 45.71 per cent, 49.54 per cent and 52.50 per cent respectively, followed by high and low categories of acquiring information from personal-localite, personal-cosmopolite and impersonal-cosmopolite sources.

Table 6.9: Distribution of respondents according to different sources of information

S.No.	*Category*		*Frequency*	*Percentage*	*Mean*	*(SD)*
A.	**Personal-localite**					
i.	Low	(<3)	60	21.43	4.36	(1.17)
ii.	Medium	(3 to 5)	128	45.71		
iii.	High	(>5)	92	32.86		
B.	**Personal-cosmopolite**					
i.	Low	(<5)	49	17.50	7.42	(2.89)
ii.	Medium	(5 to 10)	139	49.64		
iii.	High	(>10)	92	32.86		
C.	**Impersonal-cosmopolite**					
i.	Low	(<7)	48	17.14	10.02	(2.71)
ii.	Medium	(7 to 13)	147	52.50		
iii.	High	(>13)	83	29.64		

The findings depicted that the informal sources (personal-localite, personal-cosmopolite), i.e. family members, friends, progressive farmers and local leaders do not satisfy the individuals' information needs. This indicates that impersonal-cosmopolite (IT in particular) could play a pivotal role in improving access to information by the rural people. However, it remains very difficult for people with low levels of education to reap the full benefits of new technologies, including wide access to knowledge and information. These findings were in conformity with the observations of Yadav (2001), Pigato (2001) and Krikman (1999).

Extension Contact

It was observed from Table 6.10, that majority (64.64%) of the Internet users had medium extension contact, while 17.86 per cent had high and 17.50 per cent of Internet users had low extension contact (Fig. 6.2).

Table 6.10: Distribution of respondents according tc their extension contact

S.No.	*Category*		*Frequency*	*Percentage*
i.	Low	(< 8)	49	17.50
ii.	Medium	(8 to 16)	181	64.64
iii.	High	(>16)	50	17.86

Mean: 11.703 SD: 4.152

It could be understood that medium and high extension contacts are attributed due to sincere efforts of VASs, VLIs and CROs working in the study area. Still, low extension contact showed a dismal sign and this might be due to lack of follow up visits by VASs or VLIs or CROs. These findings were in conformity with the reports of Murthy (1990) and Ramchand *et al.* (1990).

Subscription of Newspaper/Magazines

Data presented in Table 6.11, indicated that 40.7 per cent of the respondents did not subscribe any paper/magazine and 59.3 per cent of the respondents subscribed to various papers / livestock

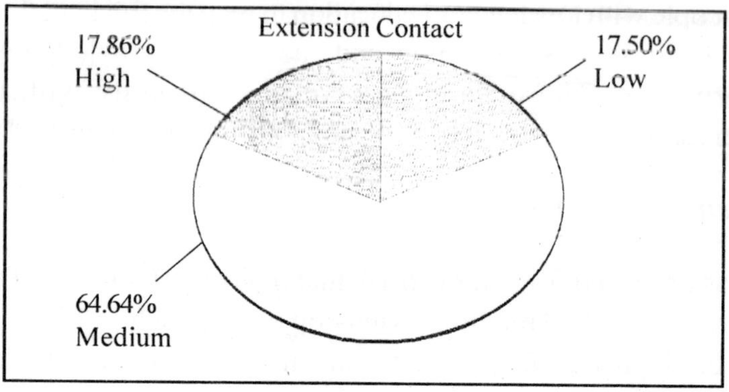

Fig. 6.2: Distribution of respondents according to their extension contact

production related magazines, namely Eenadu (61.44%), Vartha (39.16%), Annadaatha (78.31%), AgriGold Swarnasedyam (38.55%), Poultry line (20.48%) and Indian farming (12.05%).

This indicated that a fair number of farmers are interested to seek outreach information through published literatures. It is imperative that majority of these papers or magazines have online editions which can be effectively tapped by the livestock farmers.

Table 6.11: Distribution of respondents according to their subscription to news paper/magazines

S.No.	*Category*	*Frequency*	*Percentage*
A.	**Subscription**		
i.	Yes	166	59.30
ii.	No	114	40.70
B.	**Various Papers/magazines**		
i.	Eenadu	102	61.44
ii.	Vartha	65	39.16
iii.	Annadaatha	130	78.31
iv.	AgriGlod Swarnasedyam	64	38.55
v.	Poultry line	34	20.48
vi.	Indian farming	20	12.05

Inquisitiveness

Data revealed by the respondents towards sources of information, extension contact and subscription to newspapers or magazines were summed up and presented in Table 6.12.

Table 6.12: Distribution of respondents according to their inquisitive behaviour towards Internet communication.

S.No.	*Category*		*Frequency*	*Percentage*
i.	Low	(<18)	43	15.36
ii.	Medium	(18 to 36)	179	63.93
iii.	High	(>36)	58	20.71

Mean: 27.675 SD: 9.171

The data from Table 6.12 revealed that majority (63.93%) of the respondents had shown medium level inclination towards the Internet communication, while 20.71 per cent of the respondents had high level of inclination and only 15.36 per cent of the livestock farmers indicated low inclination towards Internet communication (Fig. 6.3).

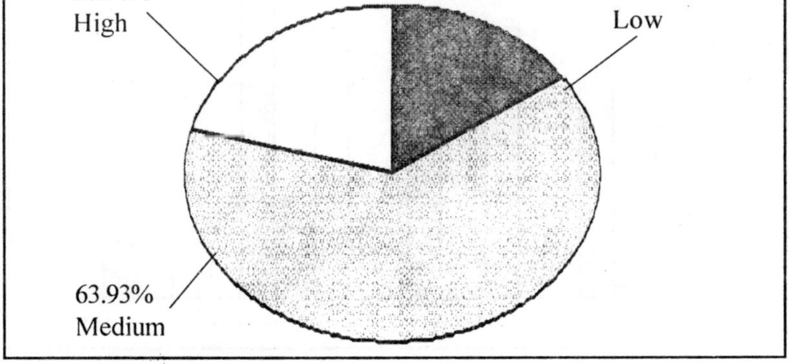

Fig. 6.3: Distribution of respondents according to their inquisitiveness

This could be due to the availability of latest information from Internet easily, within no time and with marginal cost. Similar findings were observed by Richardson (1996), Sharma (1999) and Cecchini and Scott (2003).

Internet Characteristics of Livestock Farmers

Frequency of Use

Data presented in Table 6.13 revealed that at kiosks, majority (29.30%) of the respondents used the Internet once in a week, followed by 28.20 per cent used 2-3 times in a week, 22.90 per cent used once in a month and only 3.20 per cent of the respondents accessed the Internet daily. At home, majority (8.60%) of the respondents used the Internet daily, while 7.90 per cent of the respondents used 2-3 times a week. An overall picture indicated that majority (83.60%) of the respondents connected to Internet at kiosks and only 16.40 per cent connected to Internet at home (Fig. 6.4). Which denoted the respondents preference to latest tools of ICT.

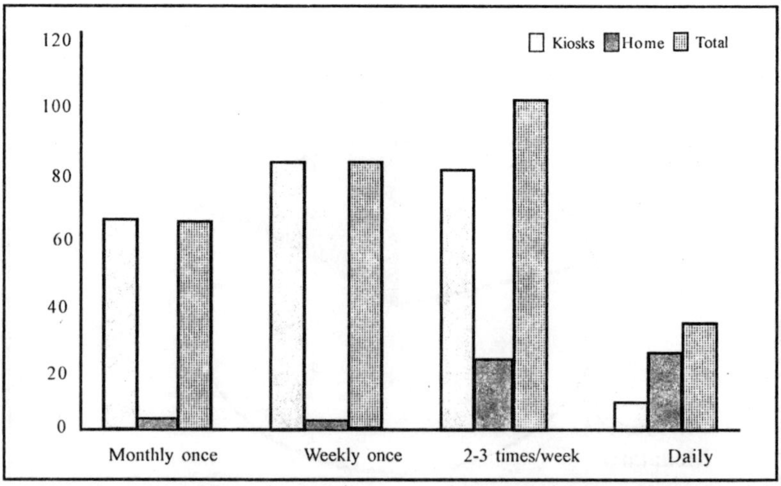

Fig. 6.4: Frequency of Internet usage by the livestock farmers

Table 6.13: Distribution of respondents according to their frequency of Internet access

S.No.	*Category*	*At Kiosks*		*At Home*		*Total*	
		F	*%*	*F*	*%*	*F*	*%*
i.	Once in month	64	22.90	0	0.00	64	22.90
ii.	Once in week	82	29.30	0	0.00	82	29.30
iii.	2-3 times/week	79	28.20	22	7.90	101	36.07
iv.	Daily	9	3.20	24	8.60	33	11.78
	Total	234	83.60	46	16.40	280	100.00

This could be due to the cost involved in installing personal computers along with Internet connection. However, the heavy benefits in poultry farming and its commerciality in nature might have motivated the farmers to go for personal computers along with Internet connection and might have encouraged the farmers to go for Internet browsing more frequently. These results were in conformity with the findings of Das (2002).

Distance Covered

Data in Table 6.14 indicate that surprisingly 10.00 per cent of the respondents visited the kiosks covering more than 10 km distance, majority (44.64%) of the respondents covered less than 2 km, while 32.15 per cent covered 2 to 5 km and 13.21 per cent of the respondents covered 6 to 10 km. of distance to access Internet (Fig. 6.5).

The above results infer that the respondents residing at distant locations visited the CICs rarely. Thus, an effort may be made to establish CICs at each village headquarter to popularise the Internet communication among all the livestock farmers.

Table 6.14: Distribution of respondents according to the distance covered by the farmers to access Internet

S.No.	*Category*	*Frequency*	*Percentage*
i.	Less than 2 km	125	44.64
ii.	2 to 5 km	90	32.15
iii.	6 to 10 km	37	13.21
iv.	Above 10 km	28	10.00

These findings were in conformity with the observations of Das (2002).

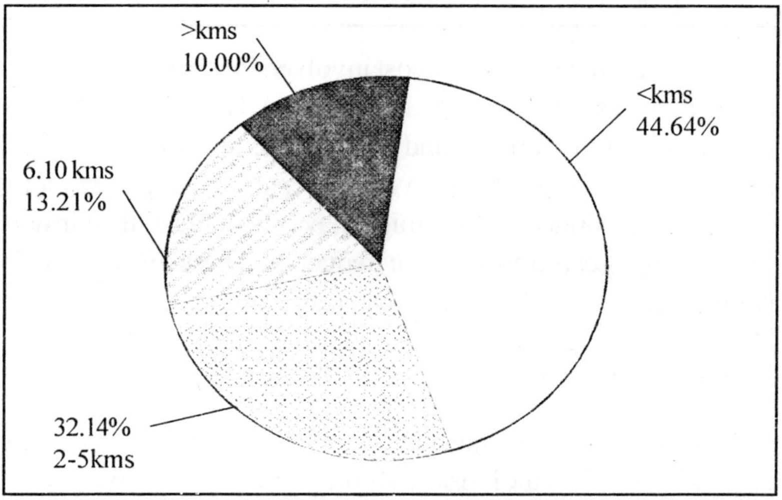

Fig. 6.5: Distance covered by the livestock farmers for Internet services

Time of Use

The information collected on time of use included the time preference, duration spent and years of usage.

Time preference: It was observed from Table 6.15, that majority (50.85% and 39.13%) of the respondents preferred afternoon hours followed by evening (32.48% and 32.61%). While 16.67 per cent and 28.26 per cent preferred morning hours at kiosks and home, respectively (Fig. 6.6).

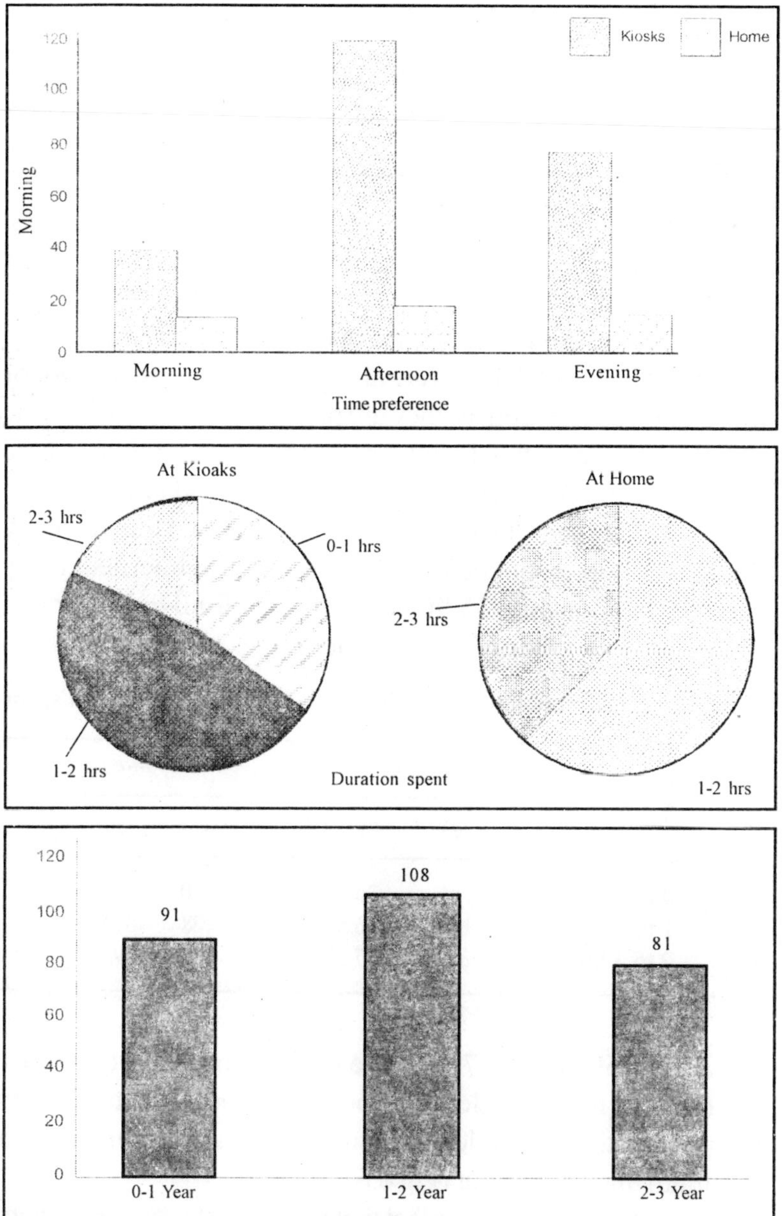

Fig. 6.6: Distribution of respondents according to their time preference, duration spent and years of usage at internet

Table 6.15: Distribution of respondents according to their time preferences

S.No.	Time preference	At Kiosks		At Home	
		F	%	F	%
i.	Morning	39	16.67	13	28.26
ii.	Afternoon	119	50.85	18	39.13
iii.	Evening	76	32.48	15	32.61

The above results testify to the due to the fact that in afternoon session the livestock farmers are able to get some leisure to go for information acquisition, as they are pre occupied in their routine animal husbandry duties during morning and evening hours. These results were in concurrence with the findings of Maniar (2002).

Duration spent: Data from Table 17 revealed that majority (46.58%) of the respondents spent their time at Internet kiosks for 1 to 2 hours, while 35.90 per cent spent for 0 to 1 hour and 17.52 per cent of them for 2 to 3 hours.

Table 6.16: Distribution of respondents according to the duration spent

S.No.	Duration spent	At Kiosks		At Home	
		F	%	F	%
i.	0 to 1 hours	84	35.90	0	0.00
ii.	1 to 2 hours	109	46.58	28	60.87
iii.	2 to 3 hours	41	17.52	18	39.13

A majority of (60.87%) of the respondents spent time for 1 to 2 hours at home while 39.13 per cent spent for 2 to 3 hours and none of the respondents spent time towards Internet for a minimum time 0 to 1 hour (Fig 6.6).

This could be due to the fact that the Internet provided variety of information at low cost and have updated information. Internet has also proved to be the great entertainer through games and movies, thus more hours spent at Internet may not be attributed to browsing of latest information. These findings were in confirmity with the observations of Maniar (2002).

Years of usage: It was revealed from Table 6.18 that majority (38.57%) of the respondents were using the Internet for the last 1 to 2 years, followed by 32.50 per cent using the Internet for less than 1 year and 28.93 per cent of the respondents were using Internet from 2 to 3 years (Fig. 6.6).

Table 6.17: Distribution of respondents according to their usage level

S.No.	*Category*	*Frequency*	*Percentage*
i.	Less than 1 year	91	32.50
ii.	1 to 2 years	108	38.57
iii.	2 to 3 years	81	28.93

The results imply that farmers who had more years of usage might have well acquainted with the new medium and might have realized the easy access and usefulness of latest information and explored the latest information. The above results were in similar lines with the observations of Maniar (2002).

Extent of Various Languages Known

Data presented in Table 6.18 that English language can be understood, speak, read and write by majority (38.58%) of the livestock owners followed by 16.79 per cent who can understand, speak and read but cannot write; 13.21 per cent who can understand and speak but cannot read and write and 6.78 per cent who could understand only. 24.64 per cent of the respondents possess no knowledge in English language. Majority (55.72%) of the livestock owners possessed no knowledge in Hindi followed by those who could understand (20.00%) only. Only 4.28 per cent of the respondents could understand, speak, read and write the Hindi language. Majority (91.43%) of the livestock owners could understand, speak, read and write in Telugu language followed by 8.57 per cent who could understand, speak and read only. Among other languages (i.e. predominantly Tamil), 22.86 per cent could understand, 18.57 per cent could understand and speak, 15.00 per cent could understand, speak and read and 6.43 per cent could understand, speak, read and write (Fig. 6.7).

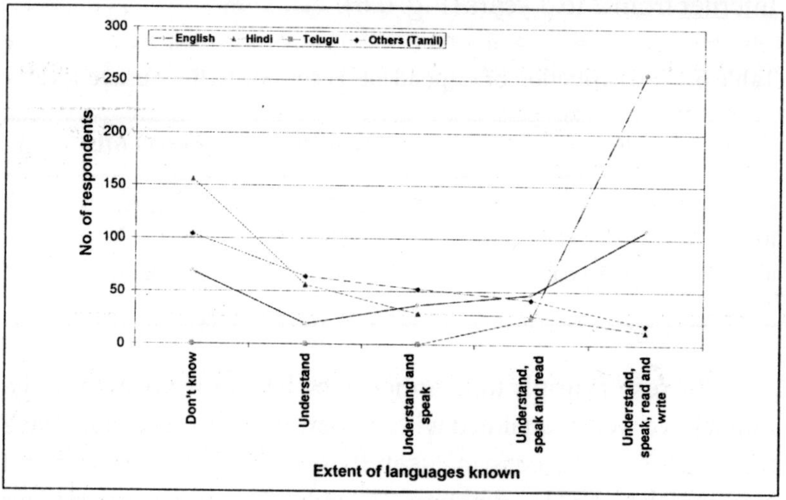

Fig. 6.7: Extent of languages known by the livestock farmers

Table 6.18: Distribution of respondents as per the extent of languages known

S.No.	*Category*	*English*		*Hindi*		*Telugu*		*Others*	
		F	*%*	*F*	*%*	*F*	*%*	*F*	*%*
i.	Don't know	69	24.64	156	55.72	0	0.00	104	37.14
ii.	Understand	19	6.78	56	20.00	0	0.00	64	22.86
iii.	Understand and speak	37	13.21	29	10.36	0	0.00	52	18.57
iv.	Understand, speak and read	47	16.79	27	9.64	24	8.57	42	15.00
v.	Understand, speak, read and write	108	38.58	12	4.28	256	91.43	18	6.43

This is due to the fact that more than 75 per cent of the respondents are educated up to higher or equal to high school, which might have helped them to understand, speak, read and write various languages including English. Since the Kuppam project is located adjacent to Tamil Nadu state, thus some of the farmers were well acquainted with Tamil language also. The above results were in similar lines with the observations of Sood (2000).

Language preference: It was clear from Table 6.19 that majority (57.50%) of the respondents preferred Telugu language to access the information through Internet and 42.50 per cent preferred English language. Whereas, none of the respondents preferred either Hindi or any other language while availing Internet services.

Table 6.19: Distribution of respondents according to their language preference

S.No.	*Category*	*Frequency*	*Percentage*
i.	English	119	42.50
ii.	Telugu	161	57.50

As more than seventy per cent of the respondents were highly educated and Telugu being the spoken language of the administration along with English, farmers might have given preference to Telugu and English languages to access the web. India stands perhaps to be the first among non-native-English speaking countries, where many people use English to access the Internet. These findings were in similar lines of observations made by Sood (2000).

Self-perception of Familiarity with Various Computer Applications

It was evident from Table 6.20 that in computer expertise, majority (46.40%) of the respondents belonged to intermediate category, followed by advanced (42.50%) and beginner (11.10%) categories. In Internet expertise, majority (41.1%) of the respondents belonged to advanced category, followed by beginner (30.00%) and intermediate (28.93%) categories. Whereas in E-mail expertise, majority (36.79%) of the respondents belonged to intermediate category, while 36.10 per cent belonged to advanced category and 27.14 per cent of the respondents belonged to beginner category.

Table 6.20: Distribution of respondents according to their expertness in various computer applications

S.No.	*Category*	*Computer*		*Internet*		*E-mail*	
		F	*%*	*F*	*%*	*F*	*%*
1.	Beginner	31	11.10	84	30.00	76	27.14
2.	Intermediate	130	46.40	81	28.93	103	36.79
3.	Advanced	119	42.50	115	41.10	101	36.10

This could be, due to the intervention of World Corps of India (Non-governmental organization) by giving training to CIC workers and converting them into most effective World Corps Professionals in addition to orientation programs to farmers in various computer applications and Internet browsing.

Table 6.21(a): Distribution or respondents according to their frequency of visits to various bilingual internet sites

S.No.	Internet sites	Regular		Often		Seldom		Never		Rank
		F	%	F	%	F	%	F	%	
1.	www.atmachittoor.com	159	56.79	81	28.93	40	14.28	0	0.00	II
2.	www.aponline.com	176	62.86	83	29.64	21	7.50	0	0.00	I
3.	www.ap-it.com	145	51.79	92	32.86	43	15.36	0	0.00	III

Table 6.21(b): Distribution or respondents according to their frequency of visits to various anilingual internet sites

S.No.	Internet sites	Regular		Often		Seldom		Never		Rank
		F	%	F	%	F	%	F	%	
1.	www.agnet.org	62	22.14	60	21.43	37	13.21	121	43.21	VIII
2.	www.agnic.org	61	22.79	46	16.43	21	7.50	152	54.28	VII
3.	www.angrau.net	108	38.57	72	25.71	45	16.07	55	19.64	IV
4.	www.google.com	129	46.07	101	36.07	50	17.86	0	0.00	I
5.	www.niir.org	117	41.79	96	23.57	37	13.21	30	10.71	II
6.	www.poultrysolutions.com	116	41.42	64	22.50	54	19.27	46	16.43	III
7.	www.meatpoultryonline.com	23	8.21	17	6.07	6	2.14	234	83.57	X
8.	www.poultry.org	26	9.29	17	6.07	3	1.07	234	82.57	IX
9.	www.poultryexhange.com	92	32.86	45	16.07	27	9.64	116	41.43	V
10.	www.vethelplineindia.com	87	31.07	66	23.57	39	13.93	90	32.14	VI

Different Internet Sites Visited by the Livestock Farmers

It was apparent from Table 6.21(a) that among bilingual sites, majority (62.86%) of the livestock owners visited www.aponline.com, which ranked first followed by www.atmachittoor.com (56.79%) and www.ap-it.com (51.79%). The figures are based on the regular visits to Internet sites. The results are depicted in Fig. 6.8 also for easy understanding.

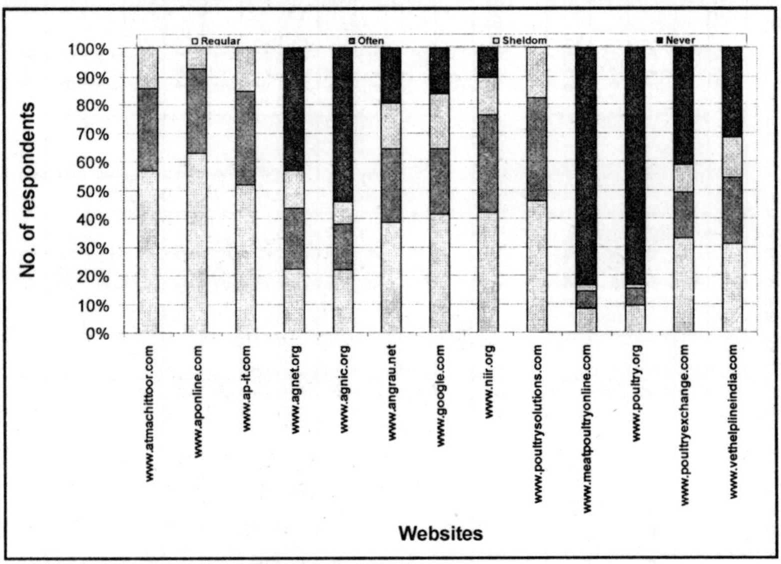

Fig. 6.8: Regularity of visits of different websites by the livestock farmers

In fact, the Andhra Pradesh state government has provided day-to-day information covering all subject areas of a common farmers' interest through its official websites viz., www.aponline.com and www.ap-it.com in local languages.

It was evident from Table 22 (b), that among anilingual sites, majority (46.07%) of the livestock owners had visited www.google.com, which ranked first followed by www.niir.org (41.79%) ranked second, www.poultrysolutions.com (41.42%)

(A) Home page of www.poultryexchange.com

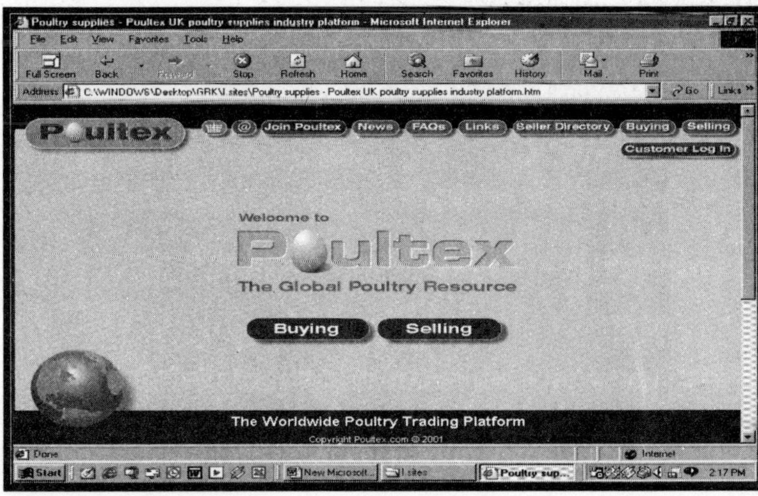

(B) Home page of www.poultrysolutions.com

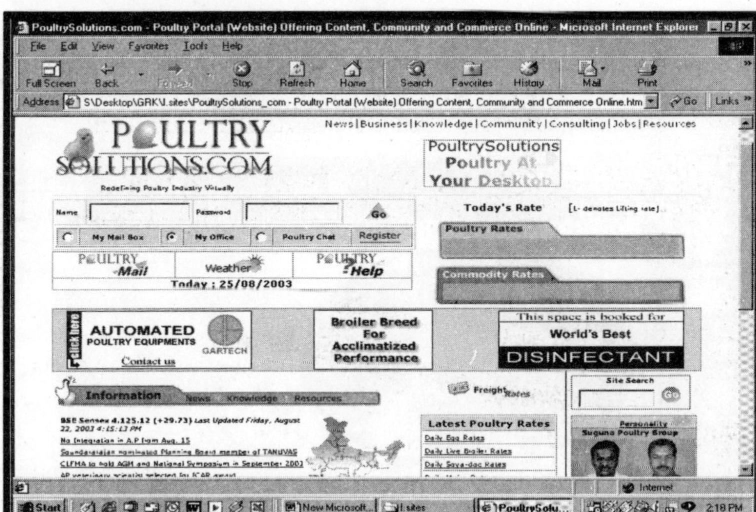

(C) Home page of www.agnet.org

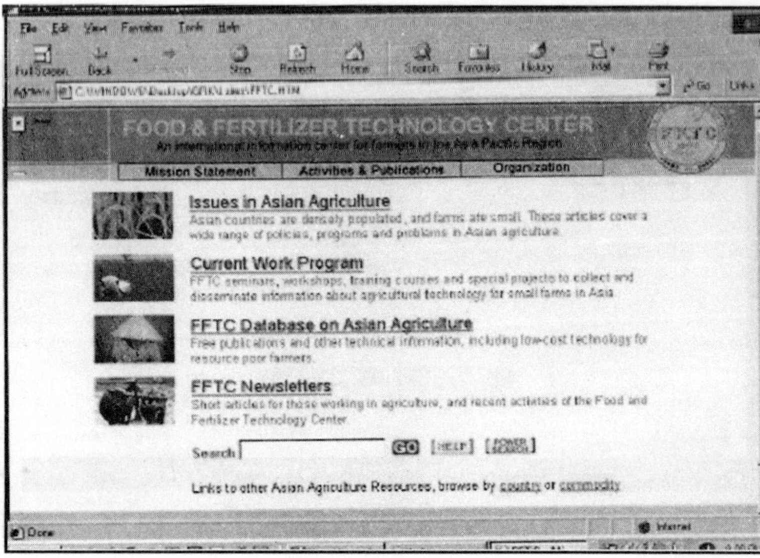

(D) Home page of www.vethelplineindia.com

(E) Home page of www.aponline.com

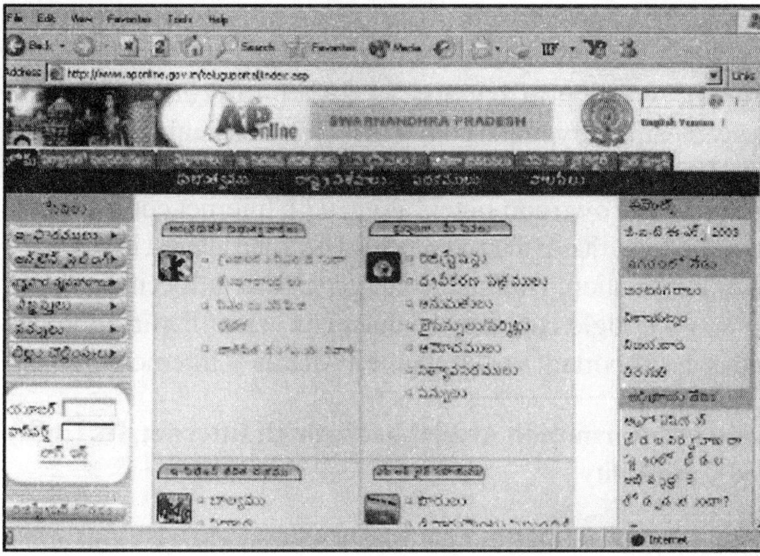

(F) Home page of www.ap-it.com

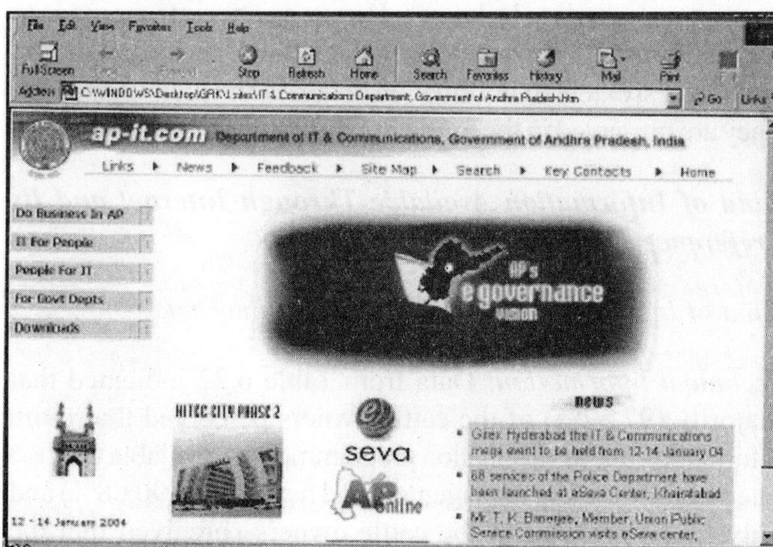

ranked third, www.angrau.net (38.57%) ranked fourth, www:poultryexchange.com (32.86%) ranked fifth, www.vethelplineindia.com (31.07%) ranked sixth, www.agnic.org (22.79%) ranked seventh, www.agnet.org (22.14%) ranked eighth, www.poultry.org (9.29%) ranked ninth and www.meatpoultryonline.com (8.21%) ranked tenth.

This is because, most of the poultry farmers and progressive farmers have own computers along with Internet connection at home and also due to their commercial bent and interest in browsing latest information. The importance given to an unspecific website, i.e. www.google.com may be due to its versatility in approach and is the important search engine available at Internet.

Kind of Information Available Through Internet, Its Utility and Credibility

Generally, every individual makes certain choices and show preferences for some areas of information over others. While making preferences, the kind of information available for farmers through web needs to be known. Hence, an attempt was made to study the kind of information available through web, farmers' preferences, extent of information obtained and their usefulness. They are presented in the following sub-heads:

Kind of Information Available Through Internet and Its Preference

Kind of Information Available Through Internet

Technical information: Data from Table 6.22 indicated that majority (97.52%) of the cattle owners perceived that more information about vaccination programme was available through Internet followed by management and health care (90.08%) and only 65.29 per cent of the cattle owners perceived that the information about breeding was available through Internet. In case

Table 6.22: Distribution of respondents according to their perception (in contrast to investigator) towards kind of information availabe through Internet and their reference (I= Investigator; F= Farmer)

S.No.	Category	*Availability of information through Internet (Frequency)*								*Prefe-rence (N=280)*
		Cattle (n=121)		*Buffalo (n=73)*		*Sheep (n=40)*		*Poultry (n=46)*		
		I	*F*	*I*	*F*	*I*	*F*	*I*	*F*	
A.	*Technical Information*									
1.	Breeding	121	79 (65.29)	73	22 (30.14)	40	15 (37.50)	46	36 (78.26)	81
2.	Feeding	121	101 (83.47)	73	31 (42.46)	40	19 (47.50)	46	40 (86.96)	146
3.	Management	121	109 (90.08)	73	45 (61.64)	40	19 (47.50)	46	43 (93.48)	153
4.	Health care	121	109 (90.08)	73	31 (42.46)	40	17 (42.50)	46	41 (89.13)	153 153
5.	Vaccination programme	121	118 (97.52)	73	31 (42.46)	40	19 (47.50)	46	43 (93.48)	181
6.	Fodder production	121	98 (80.99)	73	29 (39.73)	00	00	00	00	101
7.	Preparation of milk products	121	60 (49.59)	73	24 (32.88)	00	00	00	00	28
8.	Project proposal writing	121	102 (84.30)	73	35 (47.94)	40	15 (37.50)	46	43 (93.48)	93
	Average	121	97 (80.16)	73	31 (42.46)	40	13 (32.15)	46	41 (89.13)	117 (41.78)

B.	*Marketing Information*									
1.	Price list of medicine/ vaccine	121	91 (75.21)	73	45 (61.64)	40	11 (27.50)	46	36 (78.25)	108
2.	Sale/purchase to livestock	121	89 (73.55)	73	31 (63.01)	40	19 (27.50)	46	39 (84.73)	39
3.	Marketing of livestock products	121	91 (75.21)	73	46 (63.01)	40	11 (27.50)	46	38 (82.61)	43
4.	Current price of livestock products	121	105 (86.78)	73	58 (79.45)	40	12 (30.00)	46	40 (86.96)	191
5.	Status of national and International market	121	107 (88.43)	73	65 (89.04)	40	14 (35.00)	46	43 (93.48)	115
	Average	121	98 (80.99)	73	50 (68.49)	40	12 (30.00)	46	40 (86.96)	97 (34.64)
C	*General Information*									
1.	Rural Finance	121	99 (81.82)	73	64 (87.67)	40	29 (72.50)	46	39 (84.78)	111
2.	Insurance facilities	121	36 (29.75)	73	34 (46.57)	40	14 (35.00)	46	34 (73.91)	22
3.	Education	121	90 (74.38)	73	43 (58.90)	40	28 (70.00)	46	42 (91.30)	163
4.	Employment	121	89 (73.55)	73	44 (60.27)	40	12 (30.00)	46	41 (89.13)	116
5.	Mandi gyan	121	102 (84.30)	73	61 (83.56)	40	18 (25.00)	46	39 (45.65)	42
6.	Weather	121	109 (90.08)	73	60 (82.19)	40	18 (45.00)	46	39 (84.78)	42
7.	Government schemes	121	119 (98.35)	73	65 (89.04)	40	36 (90.00)	46	43 (93.48)	187
	Average	121	92 (76.03)	73	53 (72.60)	40	21 (52.50)	46	37 (80.43)	102 (36.46)

of buffalo owners, majority of them (61.64%) perceived that more information pertains to management was available thorough Internet followed by feeding, management and health care (42.46%), while only 30.14 per cent of the buffalo owners perceived the availability of information about breeding through Internet. Among sheep owners, majority (47.50%) perceived that the information about feeding, management and vaccination programme was more available through Internet and only 37.50 per cent of the sheep owners perceived that the information about breeding was available through Internet. In case of poultry farmers, majority of them (93.48%) perceived that information pertains to project proposal writing, vaccination programme and management was available through Internet and 78.26 per cent of the poultry farmers perceived that the information about breeding was available through Internet. Although an appreciable percentage of poultry and cattle farmers perceived that the information pertains to breeding was available through Internet but has listed last, this could be due to the lack of interest in the subject or its need in field situation.

Majority (89.13%) of the poultry owners perceived that the technical information was available through Internet followed by cattle owners (80.16%), buffalo owners (42.46%) and only 32.50 per cent of the sheep owners perceived that technical information was available through Internet (Fig 6.9). An higher percentage of poultry and cattle farmers' perception towards availability of technical information through Internet could be due to the actual interest among the farmers towards latest technical information and virtual availability of all information through Internet according to their interest.

Marketing information: It was evident from the Table 6.22 that among all the species owners, majority of them perceived that the information pertains to status of national and International market and current price of livestock products. These results imply that the livestock farmers are very much keen to know more about the surrounding market as well as the demand for their produce.

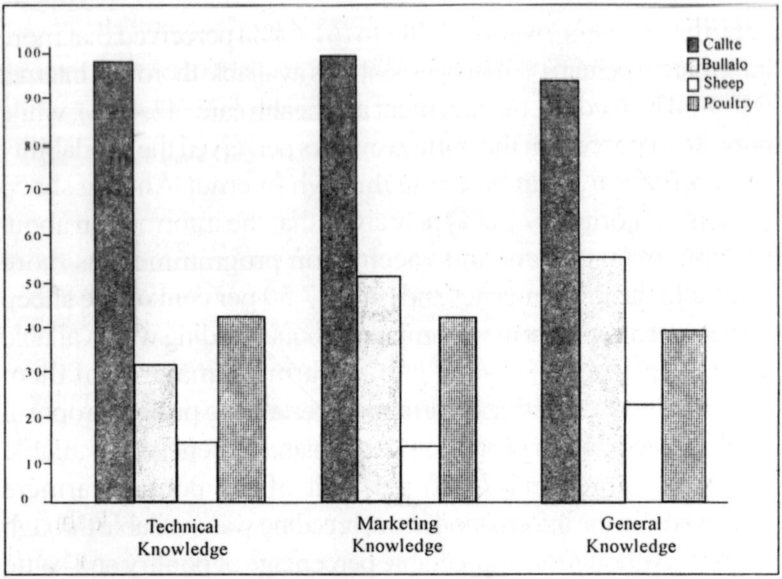

Fig. 6.9: Livestock farmers' perception towards the availability of Internet Information

Majority (86.96%) of the poultry owners perceived that the information pertains to marketing was available through Internet followed by cattle owners (80.99%), buffalo owners (68.49%) and only 30.00 per cent of the sheep owners perceived that marketing information was available through Internet (Fig 6.9). The results imply that as most of marketing information is available in numbers thus, the language and other obstacles may not discourage the farmers from obtaining this information at a greater extent. This could be due to the wide variety of consumers for poultry and dairy products, basic commercial instinct among poultry and cattle farmers and differences in individual's information needs.

General information: Data from Table 6.22 reveals that 98.35, 89.04, 35.00 and 93.48 per cent of cattle, buffalo, sheep and poultry farmers respectively perceived that the information pertains to government schemes was available through

Internet. Similarly for all the owners, except poultry owner information about insurance facility perceived to be least available through Internet. This could be due to their unawareness about insurance and its advantages.

Majority (80.43%) of the poultry owners had perceived that the general information was available through Internet followed by cattle owners (76.03%), buffalo owners (72.60%) and about 52.50 per cent of the sheep owners perceived that general information was available through Internet (Fig 6.9). A very fair percentage of all owners had perceived that general information required for their day-to-day life is available through Internet, this could be due to the farmers' general areas of interest and availability of content in local languages.

The data presented in Table 6.22 was subjected to chi-square test, to find out whether the observed frequencies (farmers' perception) are in agreement with the expected frequencies (Investigator's perception) or not. The calculated χ^2 values of various kinds of information in different species were given below:

S.No.	*Kind of information*	*Degree of freedom*	*χ^2 calculated value*			
			Cattle	*Buffalo*	*Sheep*	*Poultry*
i.	Technical	8-1=7	28.44**	19.06**	18.99**	24.54**
ii.	Marketing	5-1=4	15.21**	16.58**	14.20**	16.13**
iii.	General	7-1=6	21.33**	20.07**	19.02**	29.06**

** : $p < 0.01$

It was clear from the above results that the χ^2 calculated values among livestock owners having different species and kinds of information were greater than the χ^2 table values at one per cent level of significance ($p<0.01$).

This concludes that there is agreement among the

perceptions of investigator and farmers about kind of information available through web. This may be due to the fact that three of the websites are bilingual (i.e. both English and Telugu) and thus the respondents are able to access the websites available in Telugu and English. Further, many of the respondents have fair knowledge in English and thus, were able to access the English websites also.

Preferences: It was also evident from Table 6.22 that on an average majority (41.78%) of the livestock farmers had given their preferences to technical information followed by general information (36.43%) and 34.64 per cent of the livestock farmers had given preferences to marketing information (Fig. 6.10). This may be due to the importance of technical information in livestock production and management. The above results also infer that marketing and general information were available through other sources also in the study area.

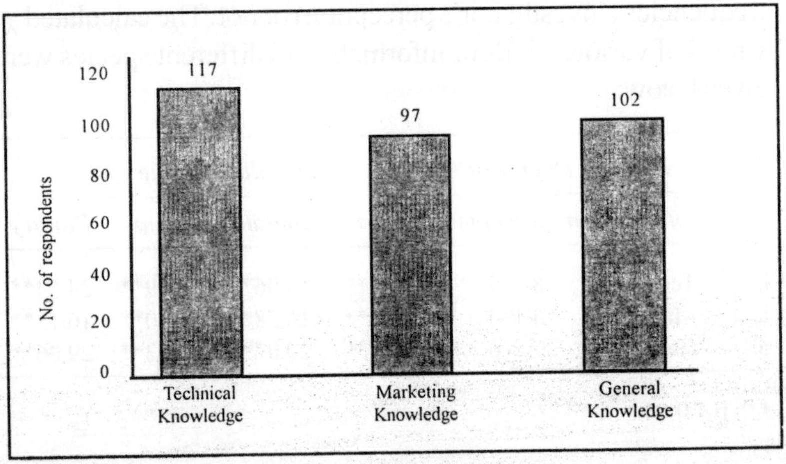

Fig. 6.10: Livestock farmers' preference about variety of Internet information

Extent of Information Obtained

Technical information: An overall view of technical information in Table 6.23 indicated that majority (55.98% and 52.71%) of livestock farmers had obtained the information to full extent pertaining to feeding and management. Whereas, only 32.14 per cent of livestock farmers had obtained full information pertains to project proposal writing (Fig. 6.11). This may be due to the fact that feeding and management were proved to be most important aspects of livestock production. While project proposal writing was not a routine practice in livestock production. The results imply non-availability of content in local language and differences in individual's information needs.

Marketing information: Data from Table 24 indicated that majority (57.45%) of livestock farmers had obtained the information about current price of livestock products to full extent, followed by status of National and International market (52.61%) and 52.04 per cent of livestock farmers obtained the information pertaining to marketing of livestock products to full extent (Fig. 6.11). The results denoted that as most of the marketing information is available in numbers thus, the language barrier may not discourage the farmers from obtaining this valuable information to a greater extent.

General information: An overall view of general information from Table 24 indicate that majority (55.55%) of the livestock farmers obtained the information pertaining to rural finance to full extent. While very low proportions (11.76% and 11.28%) of the livestock farmers obtained information pertaining to insurance facilities and employment opportunities to full extent, respectively (Fig. 6.11).

Table 6.23: Distribution of respondents as per the extent of information obtained and its usefulness

SNo	*Category*	*N*	*Extent of information Obtained (%)*			*Extent of usefulness of information (%)*			
			Full	*Some*	*Little*	*V. useful*	*Useful*	*Some*	*V. little*
A	*Technical Information*								
1	Breeding	152	31.20	64.54	4.26	45.39	30.50	17.02	7.09
2	Feeding	191	55.98	32.35	10.53	44.49	30.62	21.53	3.35
3	Management	216	52.71	35.27	12.02	39.15	43.02	15.12	27.13
4	Health care	206	42.69	37.94	19.36	39.13	42.88	16.21	1.98
5	Vaccination schedule	211	38.43	43.39	18.18	21.49	43.39	3.16	4.96
6	Fodder production	157	32.80	57.43	9.74	55.89	27.69	14.36	2.05
7	Preparation of milk produts	84	39.18	43.30	13.51	22.43	21.62	30.27	25.68
8.	Project proposal writing	195	32.14	55.10	12.76	49.49	27.55	19.89	3.06
	Average	182	40.64	46.66	12.54	39.68	33.41	17.19	9.41

Contd.....

A	*Technical Information*								
1	Price list of medicine/vaccine	183	43.83	30.24	25.92	30.24	25.92	24.07	19.75
2	Sale/purchase of livestock	170	40.25	49.35	10.40	41.55	38.96	14.29	5.19
3	Marketing of livestock products	186	52.04	30.61	17.35	47.96	32.65	16.32	3.36
4	Curent price of live stock products	215	57.47	37.83	3.72	42.91	38.43	15.29	1.59
5	Status of national and Intenational merket	229	52.61	35.07	12.31	51.06	28.72	18.62	3.06
	Average	200	49.23	36.82	13.94	42.74	32.93	17.72	6.59
C	*General Information*								
1	Rural Finance	231	55.55	34.78	8.69	43.96	31.40	21.26	3.83
2	Insurance facilities	118	11.76	48.53	39.71	33.82	22.06	22.06	19.11
3	Education	203	18.26	38.17	43.57	42.74	34.85	19.92	1.03
4	Employment	186	11.28	41.02	47.69	47.69	33.33	17.95	2.48
5	Mandi gyan	194	41.60	38.55	19.84	45.78	32.53	20.48	1.20
6	Weather	226	40.00	32.65	27.35	46.94	33.67	16.33	2.07
7	Governmental schemes	263	45.87	35.95	18.18	40.91	32.23	24.79	3.06
	Average	203	32.04	38.52	29.29	43.12	31.44	20.40	4.62

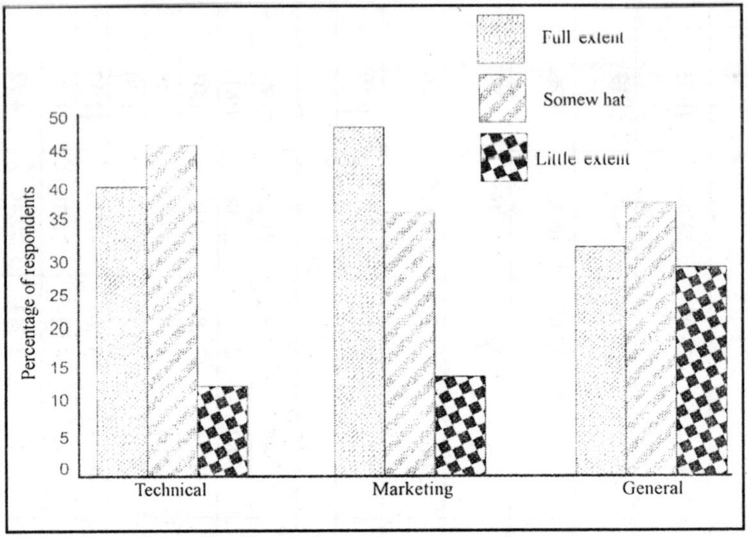

Fig. 6.11: Extent of information obtained by the livestock farmers

This could be due to the fact that small and marginal farmers are much involved in livestock production and they need an initial capital to start up with any such profit-assured enterprise. Although the availability of content in local languages can make ICT applications more accessible to poor people; gender, class and caste inequalities and differences in farmers' needs are powerful obstacles to the extent obtained.

Utility of Information

Technical information: Data from Table 6.23 indicate that majority (43.42%) of the respondents perceived the information about fodder production as very useful. While 39.15 per cent and 39.13 per cent of the respondents perceived that the information pertaining to health care and management were very useful respectively (Fig. 6.12). This infers that there is no common grazing land for livestock and thus, respondents were much interested in fodder production.

Marketing information: Table 6.23 reveal that majority (47.96%) of the respondents perceived the information about marketing of livestock products as very useful and only 30.24 per cent of the respondents revealed that the price list of medicines/ vaccines was very useful (Fig. 6.12). Very low percentage of the respondents felt that the information pertaining to marketing aspects available on web was very useful, this may be because most of the respondents are small and marginal farmers with low herd size and the non-relevancy of price list to them.

General information: Data from Table 6.23 indicate that majority (47.69%) of the livestock owners had revealed that the education related information was very useful followed by governmental schemes (46.94%), mandi gyan (45.78%), rural finance (43.96%), employment (42.74%), weather (40.91%) and only 33.82 per cent of the livestock owners had revealed that the information about insurance facilities was very useful (Fig. 6.12). In all the areas of general information, less than half of the respondents felt that the general information is very useful; this is due to the fact that there are other sources to provide general information like newspapers, television, radio and extension officer.

Change in information acquisition habits: Data from the Table 6.24a reveal that majority (60.00%) of the livestock owners had changed their habit of information acquisition after being acquainted with Internet and 40.00 per cent had not changed their information acquisition habits.

Table 6.24a. Distribution of respondents according to their change in information acquisition habits (N=280)

S.No.	*Change in information acquisition habits*	*Frequency*	*Percentage*
i.	Yes	168	60.00
ii.	No	112	40.00

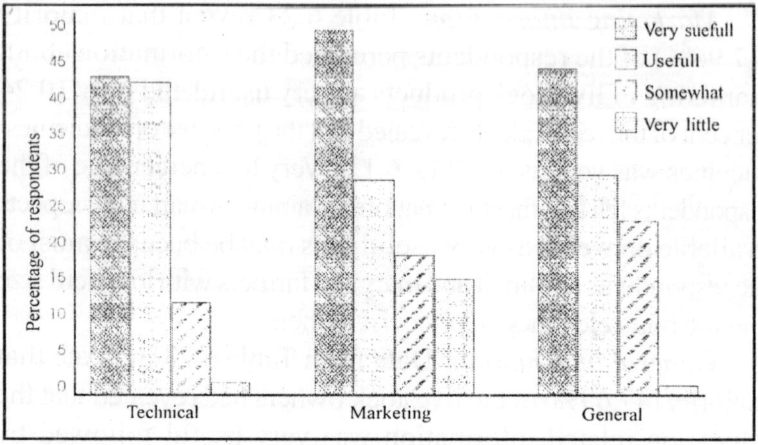

Fig. 6.12: Extent of utilization of Internet information by the livestock farmers

The above results were purely relying on individuals' perceptions, which in turn based on mode of understanding, languages known, accessibility to other sources of information etc. Similar type of observations were made by Maniar (2002).

Table 6.24b: Distribution of respondents as per the change in information acquisition habits at different frequencies (n=168)

S.No.	*Change in information acquisition habits*	*Extent of change*					
		Greater		*Some*		*Lesser*	
		F	*%*	*F*	*%*	*F*	*%*
i.	There is decrease in frequency of reading printed materials (Newspaper, magazines)	33	19.64	21	12.5	17	10.12
ii.	There is decrease in search of information from other sources (Progressive farmers, radio, Television)	14	8.33	27	16.07	16	9.52
iii.	There is reduction in frequency of visiting to Subject Matter specialist (Ext. officers, V.A.S., Univ. personnel)	8	4.76	19	11.31	13	7.74

(G) Community Information Centre, Shanthipuram

(H) Livestock farmer at Community Information Centre

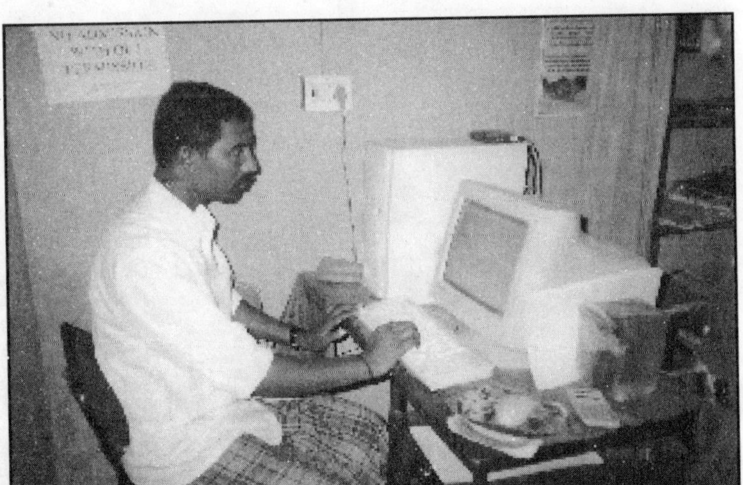

(I) Community Information Centre, Gudipalli

(J) Women farmer at Community Information Centre

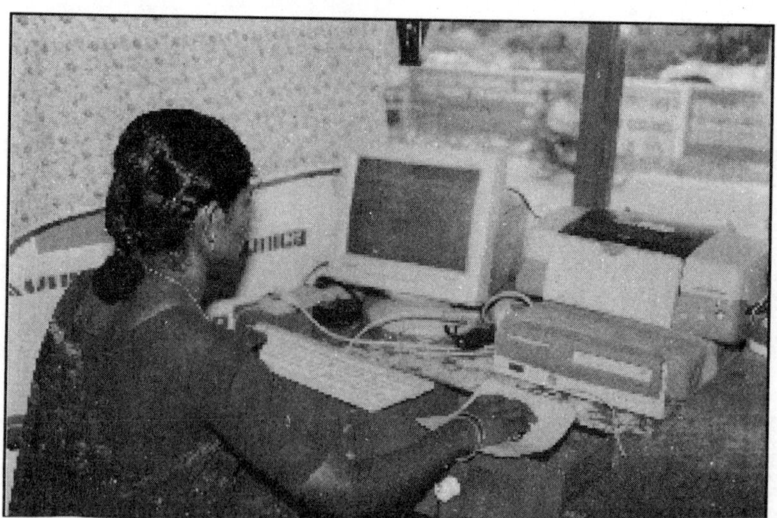

It was evident from Table 6.24b, that majority (19.64%) of respondents had decreased the frequency of reading printed materials to a greater extent, followed by some extent (12.5%) and lesser (10.12%) extent of change. While, majority (16.07%) of respondents had changed the search of information from other sources, followed by lesser extent (9.52%) and higher (8.33%) extent of change. Whereas, reduction in frequency of visiting to subject matter specialists had changed mostly (11.31%) to some extent, followed by lesser extent (7.74%) and higher (4.76%) extent of change had noticed.

This suggests that the flow of information from traditional sources like progressive farmers, extension officers, V.A.S, television, radio and newspapers do not satisfy the individuals' information needs. Whereas, the new Internet communication had ensured easy and low-cost access to latest information. This might have made the individuals to shift their habit of information acquisition from traditional to Internet sources. This had also suggested that Internet communication had helped to replace routine contacts but it does not substitute decisional contacts. The above results were in confirmity with the findings of Meier-Dallach (1998), Jung (2001) and Maniar (2002).

Credibility of Internet Sources of Information

Table 6.25 explains that 51.31 per cent of the respondents considered the Internet source as trustworthy, 49.79 per cent agreed that Internet source is free from error, 43.37 per cent agreed that Internet source serves complete and adequate information, while 33.19 per cent of the respondents agreed that Internet sources informs to a greater degree about the subject and 45.70 per cent of the respondents felt that Internet source provides latest and timely information.

Table 6.25: Distribution of respondents according to their perception about source credibility

S.No.	Statements	N	Agreement		Disagreement	
			F	%	F	%
1.	Information available through Internet is trustworthy in nature	228	117	51.31	111	48.69
2.	Information available through Internet is free from error and correct in nature	239	119	49.79	120	50.21
3.	Information available through Internet is complete and adequate in nature	249	108	43.37	141	56.63
4.	The Internet sources informs to a greater degree about a particular topic	244	81	33.19	163	66.81
5.	The Internet sources maintains and provides latest and timely information	256	117	45.70	139	54.30

The result of this study in general proves that with the current pitfalls and shortcomings the respondents availed Internet services may not be in position to create the good image and credibility that they perceive or have in their mind, there is a possibility of making Internet communication better and that would contribute to the building up of credibility. Similar observations were made by West (2000), Schweiger (2000) and Flanigan and Metzger (2000).

Knowledge and Perceived Benefits of Livestock Farmers Attained Through Internet

Extent of Knowledge

The knowledge score possessed by the respondents after acquiring knowledge through Internet and other sources about technical, marketing and general aspects are presented under four sub-heads:

Table 6.26: Distribution of respondents according to their knowledge score about technical, marketing and general information acquired through Internet and other sources

S. No.	Category	Cattle (n=121)			Buffalo (n=73)			Sheep (n=40)			Poultry (n=46)		
		Knowledge score	Through others	Through Internet	Knowledge score	Through others	Through Internet	Knowledge score	Through others	Through Internet	Knowledge score	Through others	Through Internet
A.	*Technical Information*												
1.	Breeding	0.833 (100.00)	0.665 (79.95)	0.168 (20.05)	0.500 (100.00)	0.500 (100.00)	0.000 (0.00)	0.500 (100.00)	0.500 (100.00)	0.000 (0.00)	.0750 (100.00)	0.375 (50.00)	0.375 (50.00)
2.	Feeding	0.800 (100.00)	0.600 (75.00)	0.200 (25.00)	0.700 (100.00)	0.600 (85.72)	0.100 (14.28)	0.600 (100.00)	0.600 (100.00)	0.000 (0.00)	1.000 (100.00)	0.700 (70.00)	0.300 (30.00)
3.	Management	0.875 (100.00)	0.750 (85.72)	0.125 (14.28)	0.750 (100.00)	0.650 (86.67)	0.100 (13.33)	0.633 (100.00)	0.500 (78.99)	0.133 (21.01)	1.000 (100.00)	0.650 (65.00)	0.350 (35.00)
4.	Health care	1.000 (100.00)	0.800 (80.00)	0.200 (20.00)	0.800 (100.00)	0.700 (87.50)	0.100 (12.50)	0.633 (100.00)	0.533 (84.20)	0.100 (15.80)	0.000 (100.00)	0.800 (80.00)	0.200 (20.00)
5.	Vaccination schedule	1.000 (100.00)	1.000 (100.00)	0.000 (0.000)	0.600 (100.00)	0.600 (100.00)	0.000 (0.00)	0.400 (100.00)	0.400 (100.00)	0.000 (0.00)	1.000 (100.00)	0.700 (70.00)	0.300 (30.00)
6.	Fodder production	0.600 (100.00)	0.500 (83.33)	0.100 (16.67)	0.500 (100.00)	0.400 (80.00)	0.100 (20.00)	—	—	—	—	—	—
7.	Prep. of milk products	0.333 (100.00)	0.333 (100.00)	0.000 (0.00)	0.666 (100.00)	0.666 (100.00)	0.000 (0.00)	—	—	—	—	—	—
8.	Project proposal writing	1.000 (100.00)	0.600 (7.00)	0.400 (40.00)	1.000 (100.00)	0.600 (60.00)	0.400 (40.00)	1.000 (100.00)	1.000 (100.00)	0.000 (0.00)	1.000 (100.00)	0.500 (50.00)	0.500 (50.00)
	Average	0.800 (100.00)	0.635 (78.88)	0.170 (21.12)	0.686 (100.00)	0.585 (85.28)	0.101 (14.72)	0.628 (100.00)	0.89 (93.79)	0.039 (6.21)	0.958 (100.00)	0.631 (60.65)	0.337 (39.35)
A.	Markeing information	0.714 (100.00)	0.539 (78.13)	0.175 (21.87)	0.600 (100.00)	0.500 (83.33)	0.100 (16.67)	0.400 (100.00)	0.350 (87.50)	0.050 (12.50)	0.850 (100.00)	0.425 (50.00)	0.425 (50.00)
B.	General information	0.714 (100.00)	0.428 (59.94)	0.286 (40.06)	0.571 (100.00)	0.478 (83.71)	0.093 (16.29)	0.428 (100.00)	0.378 (88.32)	0.050 (11.68)	0.800 (100.00)	0.500 (62.50)	0.300 (37.50)
	Grand-total	0.744 (100.00)	0.534 (72.85)	0.202 (27.15)	0.619 (100.00)	0.521 (84.17)	0.098 (15.83)	0.485 (100.00)	0.439 (90.52)	0.046 (9.48)	0.859 (100.00)	0.515 (59.26)	0.354 (40.74)

Technical information: It was apparent from Table 6.26 that the cattle owners possessed cent percent knowledge (score: 1.000) in health care, vaccination programme and project proposal writing and obtained 0.875, 0.833 and 0.800 knowledge score in management, breeding, feeding and fodder production respectively. Only 0.333 knowledge score about preparation of milk products was obtained by the cattle owners. The cattle owners attained highest (40.00%) of the knowledge score pertaining to project proposal writing through Internet followed by feeding (25.00%), health care (20.00%), management (14.28%) and only 16.67 per cent of the knowledge score about fodder production by gaining knowledge through Internet. While the cattle owners had not attained any knowledge through Internet about breeding, vaccination programme and preparation of milk products.

The results infer that the cattle owners had higher knowledge about various technical aspects of livestock production. Knowledge is the basis to take up any innovation. The higher knowledge among the respondents may be due to their higher levels of literacy, herd size and inquisitiveness towards latest form of communication technologies.

Data from Table 6.26 reveal that the buffalo owners had shown similar trend of that cattle owners, but had acquired a less per cent of knowledge through Internet about buffalo production in the areas of fodder production (20.00%), feeding (14.18%), management (13.33%) and health care (12.50%). The above results imply that the available technical literature about buffalo production through Internet either may not be sufficient or may not be relevant to the local conditions.

In case of sheep owners, though they had possessed a fair knowledge, but had not acquired any percentage of knowledge through Internet about sheep production except 21.01 per cent of management and 15.80 per cent of health care. Less number of sheep farmers and their low herd size might have changed their areas of interest in sheep production and the results also infer that either there is no enough websites for sheep farmers or non-availability of information in local languages.

The data in Table 6.26 reveal that the poultry owners had possessed cent-per-cent knowledge score (1.000) in all the areas of poultry production. This could be due to the poultry farmers' positive approach, enormous interest, good marketing orientation and industrial bent of mind.

While, the poultry owners had attained highest (50.00%) of the knowledge score pertaining to breeding and project proposal writing followed by management (35.00%), feeding and vaccination programme (30.00%) and 20.00 per cent of the knowledge score about health care by acquiring knowledge through Internet. To withhold the increased production in the flock and to be there in the market of national as well as of international levels, it is necessary to brush up individual's knowledge daily with latest and timely information. Believing these principles, the poultry farmers might have perceived that the available literature about poultry production through Internet is latest, timely and trustworthy.

An overall view of technical information indicates that all the owners possessed very fair knowledge in their respective fields. While the poultry owners had attained high (39.35%) of technical knowledge through Internet followed by cattle owners (21.12%), buffalo owners (14.72%) and only 6.21 per cent of the technical knowledge was attained through Internet among sheep farmers (Fig. 6.13). The higher percentage of knowledge attained through Internet among poultry and cattle owners may be due to their higher levels of literacy, herd size and inquisitiveness towards Internet technologies.

Marketing information: It was clear from the Table 6.26 that the poultry owners had revealed higher knowledge (score: 0.850) pertaining to marketing information followed by cattle owners (0.714), buffalo owners (0.600) and the sheep farmers possessed a low knowledge score (0.400). The poultry farmers obtained more (50.00%) of marketing information through Internet

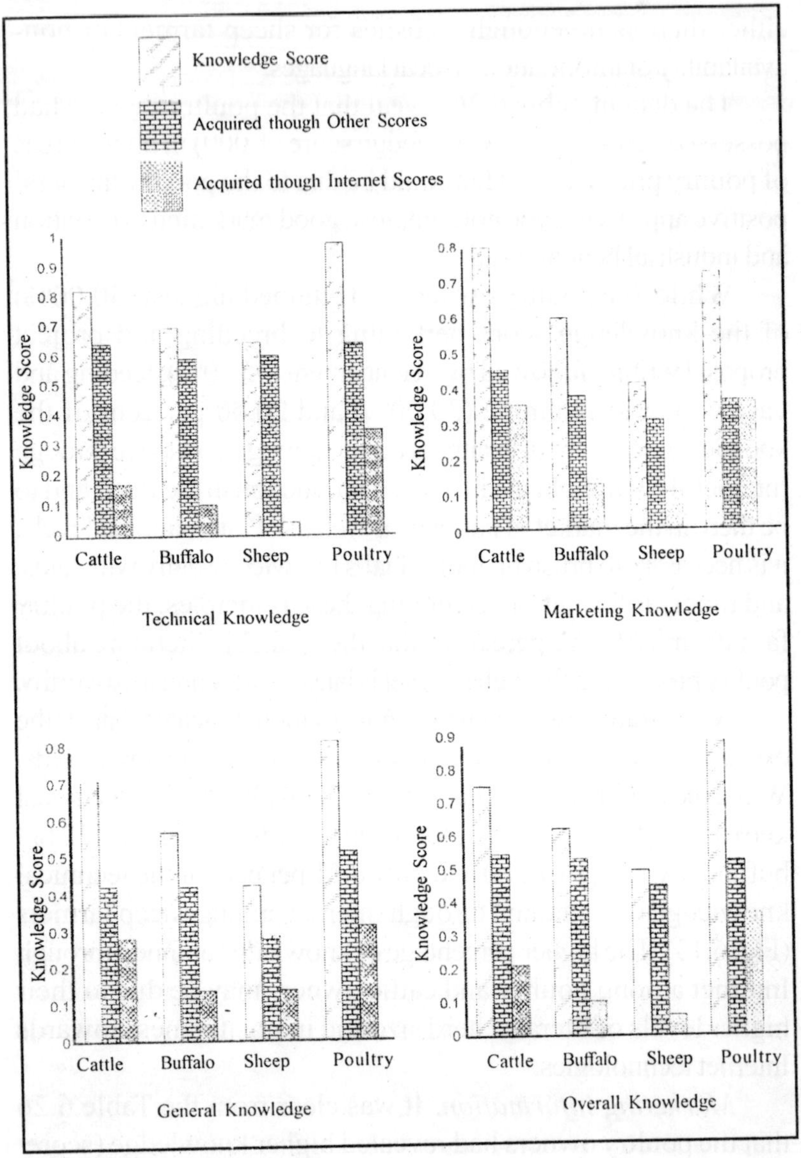

Fig. 6.13: Distribution of respondents according to their knowledge possession and acquisition through different sources

followed by cattle owners (21.87%). While buffalo owners had obtained 16.67 per cent of marketing related information through Internet. Only 12.50 per cent of the marketing information was obtained through Internet by the sheep farmers (Fig. 6.13).

Knowledge about marketing information will influence the profitability in livestock farming. So it is a good sign that the respondents had possessed higher knowledge. However, there is need to provide more and more marketing related information such as price list of medicines or vaccines or livestock products and status of National and international market through websites especially in local languages to improve the knowledge level of the farmers.

General information: It was evident from Table 6.26 that the poultry owners had higher knowledge (score: 0.800) pertaining to general information followed by cattle (0.714), buffalo owners (0.571) and the sheep farmers possessed a low knowledge score (0.428). The cattle farmers had attained more (40.06%) of general information through Internet followed by poultry owners (37.50%). While buffalo owners had obtained 16.29 per cent of general information through Internet. Only 11.68 per cent of the general information was attained through Internet by the sheep farmers (Fig. 6.13).

Knowledge is the basic requirement for human development. The above results confirm that all the owners had very fair knowledge score regarding general information. This may be due to the easy access to Internet, relevancy of content to the farmers' conditions, information availability in local languages to the livestock farmers and micro level intervention of Andhra Pradesh State Government in rural development through Internet and other ICT technologies.

Knowledge gained (Exclusively through Internet Sources)

The information attained exclusively through Internet for the first time by the livestock farmers were identified and given in Table 6.27.

Table 6.27: Distribution of respondents according to their exclusive knowledge gain through Internet sources

S.No.	*Knowledge gained*	*Frequency*	*Percentage*
1.	Different methods of breeding	18	6.43
2.	Various unconventional feeds	21	7.50
3.	Different feed additives	11	3.93
4.	Diet manipulation to change egg-yolk colour	9	3.21

It was evident from Table 6.27 that a considerable number of livestock farmers had attained different aspects of technical knowledge exclusively through Internet sources. Some (7.50%) of the respondents attained information about various unconventional feeds, viz. feeding mango kernal cake and tapioca meal. 6.43 per cent of the respondents attained knowledge about different methods of breeding, viz. In-breeding, and out-breeding. While 3.93 per cent of the respondents gained information about different feed additives viz., Prebiotics (Polysacchrides and Oligosacchrides), Probiotics (Yeast culture and Lactobacilli culture) and Antibiotics through Internet. Only 3.21 per cent of the respondents attained knowledge pertaining to diet manipulation to change egg-yolk colour through Internet sources, i.e. by including yolk colourants or pigments in poultry ration by feeding ingredients such as yellow maize, algae, dried tomato pulp, marigold petal meal and dried goat waste etc.

Analysis of Variance

The significant difference among small, medium and large categories of farmers as well as cattle, buffalo, sheep and poultry farmers in respect of knowledge were dealt by employing one-way analysis of variance.

Table 6.28: ANOVA between different categories of farmers and their knowledge level

Source of variation	*DF*	*SS*	*MS*	*'F' value*
Categories	3-1=2	411.11	205.55	41.78**
Error	280-3=277	1362.84	4.92	
Total	280-1=279			

** $p<0.01$

The observed value of 'F' is greater than the 'F' table value for 2 and 277 degrees of freedom proving thereby, that there is a significant ($p<0.01$) difference among the three categories of farmers pertaining to knowledge level. Further more, the critical difference (CD) and mean difference (MD) of these three categories of farmers showed that MD was lower than CD in respect of small-medium and small-large farmers, concluded that the extent of knowledge among these three categories of farmers were not statistically significant. This may be due to the uniformity of respondents with respect to higher levels of literacy, motivation and commerciality involved in livestock farming.

Table 6.29: ANOVA between different species owners and their knowledge level

Source of variation	*DF*	*SS*	*MS*	*'F' value*
Categories	4-1=3	1231.08	410.36	37.17**
Error	280-4=276	3047.04	11.04	
Total	280-1=279			

** $p<0.01$

The observed value of 'F' is greater than the 'F' table value for 3 and 276 degrees of freedom proving thereby, that there is a significant ($p<0.01$) difference among the four species owners about knowledge level. This could be due to the difference among the species owners with respect to literacy levels, herd/flock size and market value and commerciality involved in different livestock farming.

Table 6.30: Comparison between kiosks and home (place of access) with respect to knowledge

S.No.	*Variable*	*Mean score*		*Standard deviation*		*'Z' value*
		Kiosks	*Home*	*Kiosks*	*Home*	
1	Knowledge	11.75	9.17	3.16	3.15	4.45**

** $p<0.01$

The perusal of Table 6.30 reveals that a significant 'Z' value in terms of knowledge gained between kiosks and home as place of access. The higher knowledge among those respondents, who avail Internet services at home in comparison to kiosks, may be due to easy accessibility during anytime and higher frequency of utilization of Internet.

Perceived Benefits

The social, economic and cultural benefits attained by the respondents through Internet communication as perceived by cattle and buffalo, sheep and poultry farmers, were presented in Table 6.31, 6.32 and 6.33, respectively.

Table 6.31: Distribution of cattle and buffalo owners according to their perceived benefits (n=194)

S.No.	*Statement(s)*	*Frequency*	*Percentage*	*Rank*
1.	Internet has made access to high-speed communication	189	97.42	I
2.	Internet can be used to meet the information needs of the community	178	91.75	II
3.	Economic development opportunities are made effective through Internet communication	164	84.54	III
4.	Internet communication makes an individual conversant	156	80.41	IV
5.	Internet communication has made convergence of animal husbandry technologies	144	74.23	V
6.	Employment opportunities are made possible through Internet communication	123	63.40	VI

It was clear from the above Table 6.31 that cattle and buffalo owners perceived most of the benefits through Internet. From among them most (97.42%) of the respondents perceived that Internet has made access to high-speed communication and this was ranked first. This may be, because, within no time one can communicate to other without considering physical and geographical barriers and entire globe will be available with a mouse click.

It could be further seen that 91.75 per cent of the respondents perceived that Internet can be used to meet the information needs of the community and ranked second. This is because the Internet communication through an integrated approach proved to be effective in achieving agricultural and rural development in particular and community development in general.

Similarly, 84.54 per cent of the respondents perceived that the economic development opportunities are made effective through Internet communication and ranked third. This may be due to the fact that the Internet provides latest technical and marketing information, which in turn provided economic development opportunities to the livestock farmers.

Further, 80.41per cent of the livestock owners perceived that Internet communication makes an individual conversant and ranked fourth. This is because, through Internet individual can acquire all sort of information in a deeper fashion at one place.

About three-fourth of the respondents (74.23%) perceived that Internet communication has made convergence of animal husbandry technologies and had given last but one rank. This may be due to the availability of entire information about animal husbandry technologies and its marketing aspects easily, within no time and at one place.

Nearly two-third of the cattle and buffalo owners (63.40%) perceived that employment opportunities are made possible through Internet communication and ranked last. This is due to wide-openings in different areas with minimum qualifications were available through web, that had proved to be most beneficial for unemployed rural youth.

Table 6.33: Distribution of sheep owners according to their perceived benefits (n=40)

S.No.	*Statement(s)*	*Frequency*	*Percentage*	*Rank*
1.	Employment opportunities are made possible through Internet communication	17	42.50	II
2.	Internet has made access to high speed communication	33	82.50	I

It was evident from Table 6.32 that sheep owners had not explored this latest Internet technology to the full extent. From among the sheep owners, majority (82.50%) perceived that Internet has made access to high-speed communication and ranked first. This may be due to the fact that Internet has converted the entire world into global village and providing up-to-date information and services well in advance.

Similarly, 42.50 per cent of the sheep farmers perceived that employment opportunities are made possible through Internet communication and ranked second. This could be due to the availability of wide variety of openings in different fields through web, which in turn has proved to be most beneficial for educated and unemployed youth in the rural areas.

It was apparent from Table 6.33 that poultry owners had attained maximum benefits through Internet.

Table 6.33: Distribution of poultry owners according to their perceived benefits (n=46)

S.No.	*Statement(s)*	*F*	*%*	*Rank*
1.	Internet communication has created demand for poultry products	43	93.48	I
2.	Internet has made access to high-speed communication	41	89.13	II
3.	Consumer oriented farming has made possibl through Internet	39	84.78	III
4.	Commercial and economic development opportunities are made effective through Internet communication	38	82.61	IV
5.	Internet communication has certainly overruled the middlemenship	36	78.26	V
6.	Internet communication has converted the world into global village and kept the global market under one umbrella	31	67.39	VI
7.	Internet can be used to meet the information needs of the community	21	45.65	VII

From among the benefits attained through Internet, most (93.48%) of the respondents had perceived that Internet communication has created demand for poultry products and ranked first. This may be due to availability of current price list of poultry products, which has increased sales as well as demand.

Similarly, 89.13 per cent of poultry farmers had perceived that Internet has made access to high-speed communication and ranked second. This may be because, within no time one can communicate to other without considering physical and geographical barriers and entire globe will be available with a mouse click.

Further, 84.78 per cent of the poultry farmers had perceived that consumer oriented farming has made possible through Internet and ranked third. This may be due to the fact that studies/surveys on consumer preferences and market demand available through Internet has made possible.

Majority of the respondents (82.61%) were of the opinion that commercial and economic development opportunities are made effective through Internet communication and ranked fourth. This may be due to the fact that the Internet provides latest technical and marketing information which in turn provided economic development opportunities to the poultry farmers.

More than three-fourth of the respondents (78.26%) were of the opinion that Internet communication has certainly overruled the middlemenship and ranked fifth. This is due to non-indulgence of third party either during selling or purchasing of commodities.

More than two-third of the respondents (67.39%) perceived that Internet communication has converted the world into global village and kept the global market under one umbrella and ranked last but one. This may be due to the fact that without considering physical and geographical barriers and entire information at global level will be available within no time.

Further, 45.65 per cent of the poultry owners were of the opinion that Internet can be used to meet the information needs of the community and ranked last. This is because, the Internet communication through an integrated approach had proved to be effective in achieving agricultural and rural development and in general the community development.

Relationship Between Dependent and Independent Variables

The pearson's coefficient of correlation was worked out to find out the relationship between personal and Internet characteristics of the respondents with their knowledge level and perceived benefits. For easy understanding, this section of the chapter was presented under the following sub-heads:

Relationship Between Knowledge and Independent Variables

It was apparent from Table 6.34 that the correlation between extent of knowledge and independent variables had shown similar trend among all species owners (Fig. 6.14). The results are explained as below:

Table 6.34: Correlation coefficients ('r' values) between extent of knowledge and independent variables

S.No.	*Indepndent Variables*	*'r' value (Cattle & buffalo farmers)*	*'r' value (Sheep farmers)*	*'r' value (Poultry farmers)*
1.	Age	-0.2806 *	-0.2444 *	-0.2276 *
2.	Gender	0.1456 NS	0.1906 NS	0.1777 NS
3.	Socio-economic status	0.5456 **	0.3911 **	0.4107 **
4.	Herd size	0.5219 **	0.3567 **	0.4007 **
5.	Total Income	0.5696 **	0.3311 **	0.4132 **
6.	Language	0.3077 *	0.2899 *	0.2906 *
7.	Extension contact	0.3581 **	0.3767 **	0.2928 **
8.	Inquisitiveness	0.3492 **	0.3119 **	0.3092 **
9.	Type of information	0.1964 NS	0.1114 NS	0.1847 NS
10.	Time of use	0.2115 NS	0.2101 NS	0.1664 NS
11.	Frequency of use	0.3919 **	0.3664 **	0.3957 **
12.	Distance covered	0.1299 NS	0.1168 NS	0.1175 NS
13.	Utility of information	0.4254 **	0.3765 **	0.3156 **
14.	Credibility	0.3818 **	0.3611 **	0.6271 **

NS : Non-significant; * : $p < 0.05$; ** : $p < 0.01$

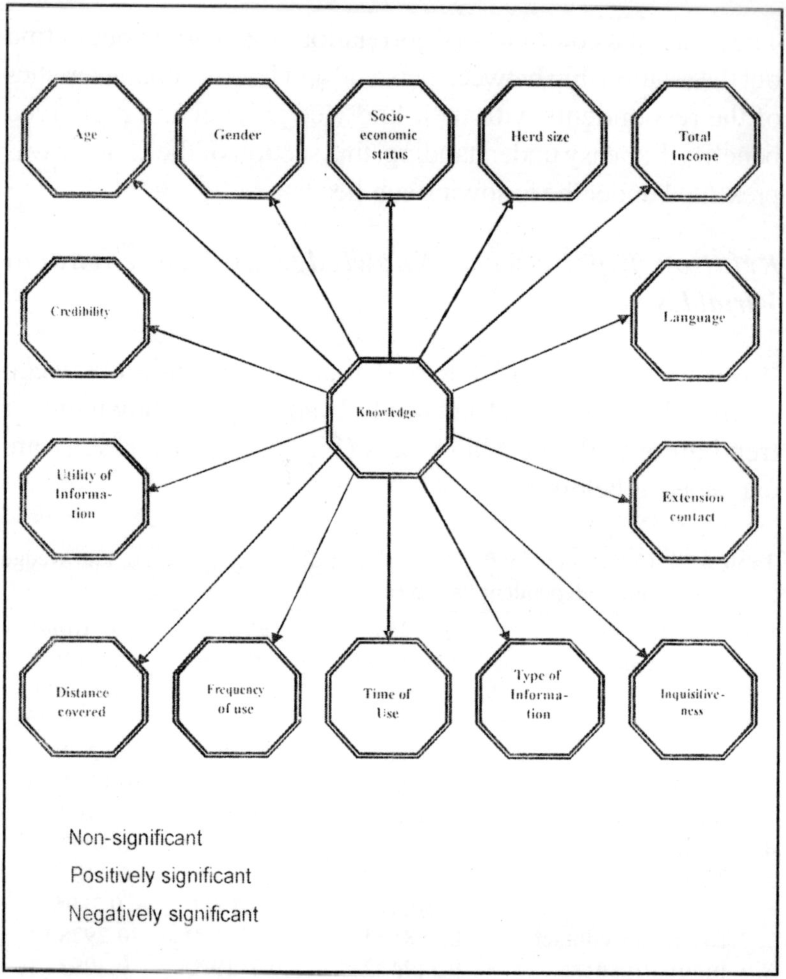

Fig. 6.14: Relationship between Knowledge and Independent Variables

Age was found to be negatively and significantly ($p<0.05$) correlated with knowledge of respondents, which may be due to more enthusiasm in knowing new technologies as reflected by higher literacy level among younger people than adult and old. Young age respondents mould their action by engaging themselves with physical and social environment around him and consequential changes takes place. More dynamism and scientific orientation of young respondents could be the reason for present outcome. Similar type of results were observed by Ruth (2000).

Socio-economic status was found to be positively and significantly ($p<0.01$) correlated with knowledge of livestock farmers. This is because, in rural areas socio-economic status is the determining factor and thus higher and middle order status groups can utilize the opportunities to a maximum extent. These results were in conformity with the findings of Hiltz and Johnson (1999) and Ruth (2000).

Herd size was positively and significantly ($p<0.01$) correlated with the knowledge of respondents. This may be due to the fact that majority of the respondents revolved around medium category of herd size and have possessed livestock farming as their main occupation.

Total income was positively and significantly ($p<0.01$) correlated with knowledge. This may be due to higher income group and also due to their larger farm and herd size that motivated the individuals to go for new communication technologies or seek latest information to improve production.

Language was positively and significantly ($p<0.05$) correlated with knowledge. This could be due to the availability of websites as anilingual or bilingual or multilingual as reflected by higher literacy level. These results were in similar line with the observations of Sood (2000).

Extension contact and inquisitiveness were found to be positively and significantly ($p<0.01$) correlated with knowledge of livestock farmers. This could be due to higher-level contacts to various sources of information as reflected by sincere efforts of V.A.S, V.L.I and local leaders and easy access to latest information through Internet. These results were in line with the findings of Richardson (1996) and Cecchini and Scott (2003).

Gender, type of information, time of use and distance covered were found to be positively and non-significantly correlated with knowledge of livestock farmers. This may be due to the uniformity of respondents in knowledge with respect to gender, type of information available through Internet, time of use and distance covered by the livestock farmers.

Frequency of use and utility of information were found to be positively and significantly ($p<0.01$) correlated with knowledge. This may be due to the fact that the increased number of exposures and extent of usage levels directly influence the knowledge possession. These observations were in coincidence with the findings of Das (2002).

Credibility was positively and significantly ($p<0.01$) correlated with knowledge of livestock farmers. This may be due to the trustworthiness, latest and complete availability information and other services that respond to the most pressing needs of the poor. These results were in line with the findings of West (2000).

Relationship Between Perceived Benefits and Independent Variables

It was evident from Table 6.35 that the correlation between perceived benefits and independent variables had shown similar trend among all species owners (Fig. 6.15). The results are expressed as below:

Age was found to be non-significant though positively correlated with perceived benefits among respondents which may be due to uniformity of respondents in knowledge with respect to age.

Socio-economic status, herd size and total income were found to be positively and significantly ($p<0.01$) correlated with perceived benefits among livestock farmers. This may be due to the fact that in rural areas socio-economic status, herd size and income gain are the determining factor either to utilize the latest opportunities or to avail benefits.

Table 6.35: Correlation coefficients ('r' values) between perceived benefits and independent variables.

S.No.	*Indepndent Variables*	*'r' value (Cattle & buffalo farmers)*	*'r' value (Sheep farmers)*	*'r' value (Poultry farmers)*
1.	Age	0.1681 NS	0.1598 NS	0.1664 NS
2.	Gender	0.1811 NS	0.1906 NS	0.1794 NS
3.	Socio-economic status	0.3103 **	0.3111 **	0.2946 **
4.	Herd size	0.3914 **	0.2967 **	0.5011 **
5.	Total Income	0.7011 **	0.3101 **	0.6502 **
6.	Language	0.2596 **	0.2799 *	0.2877 **
7.	Extension contact	0.3409 **	0.3567 **	0.3814 **
8.	Inquisitiveness	0.7112 **	0.3019 **	0.6908 **
9.	Type of information	0.1107 NS	0.1004 NS	0.2133 NS
10.	Time of use	0.1718 NS	0.2101 NS	0.2123 NS
11.	Frequency of use	0.2907 *	0.2464 *	0.2730 *
12.	Distance covered	0.1636 NS	0.1177 NS	0.0900 NS
13.	Utility of information	0.2603 *	0.3665 **	0.2701 *
14.	Credibility	0.2411 *	0.2409 *	0.2599 *

NS : Non-significant; * : $p < 0.05$; ** : $p < 0.01$

Language was positively and significantly ($p<0.05$) correlated with perceived benefits. This may be due to the multilingual access to Internet as reflected by higher literacy, will enable the farmers to access much more information available in different languages. These observations were in conformity with the findings of Sood (2000).

Extension contact and inquisitiveness were found to be positively and significantly ($p<0.01$) correlated with perceived benefits among livestock farmers. This could be due to higher

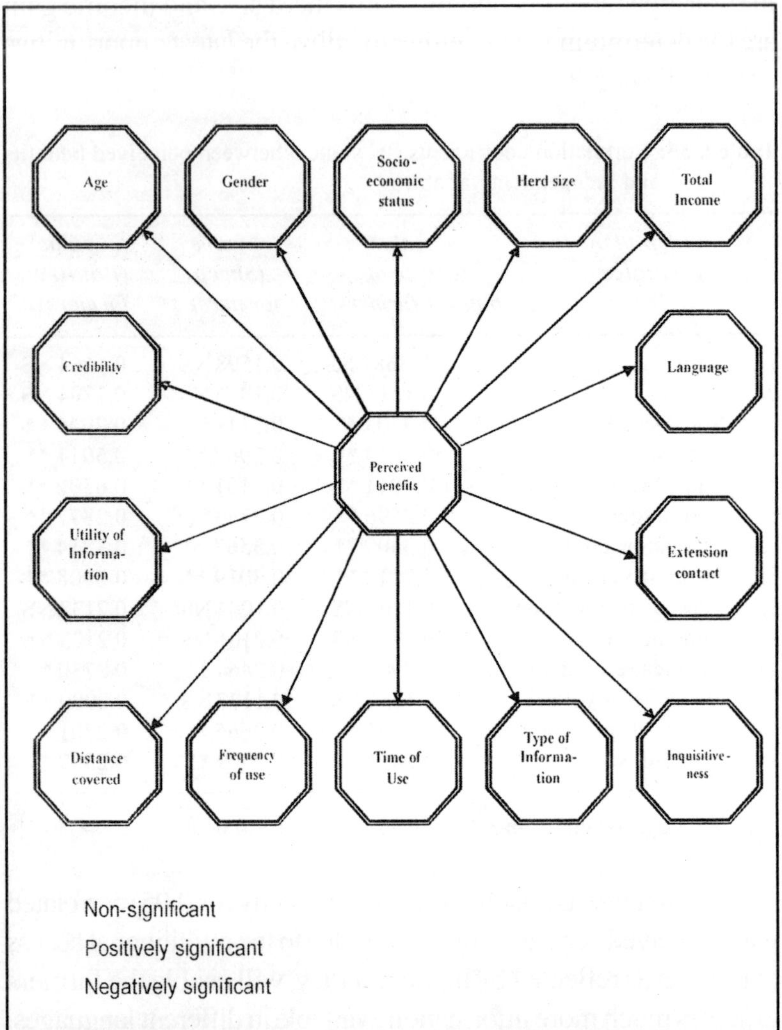

Fig. 6.15: Relationship between perceived benefits and Independent Variables

level contacts to various sources of information as reflected by sincere efforts of V.A.S, V.L.I and local leaders and easy access to latest information through Internet and avail benefits.

Gender, type of information, time of use and distance covered were found to be positively but non-significantly correlated with perceived benefits. This may be due to the uniformity of respondents in perception of advantages or benefits with respect to gender, type of information available through Internet, time of use and distance covered by the livestock farmers.

Frequency of use and utility of information were found to be positively and significantly ($p<0.05$) correlated with perceived benefits. This may be attributed to low-cost and easy accessibility of Internet, as well as practicality, need and demand of information available through Internet.

Credibility was positively and significantly ($p<0.01$) correlated with perceived benefits. This may be due to the fact that the information through web is accurate, complete, fair and trustworthy. These results were in conformity with the findings of West (2000).

The perceptions of livestock famers are more likely influenced by their personal experience and suggestions from the Internet users. The farming communities generally live in rural areas with low standards of living and their caste structure, creed composition, group affiliation and socio-economic status will exert utmost influence on their perception. The people who avail Internet services in rural areas are also considered to have similar perceptions with regard to IT benefits as in case of people living in urban areas.

Factors Influencing the Perceived Benefits of Livestock Farmers

Apart from independent variables, certain other factors were considered to study their influence on perceived benefits, such as categories of livestock farmers, place of Information access, computer literacy and level of information intensity.

Categories of farmers: The correlation if any among the three categories of farmers (Small, medium and large) based on herd size and perceived benefits of respondents were dealt by employing coefficient of concordance. All the respondents as well as their perceived benefits were combined and depicted in Table 6.36 to arrive at the following results.

The calculated value of coefficient of concordance (w) was 0.621, this positive correlation indicated that the farmers' perception towards the benefits attained through Internet was not differed with their categories. The three categories of farmers stand in same direction so far as their perceptions towards the benefits attained through Internet. These are the results due to higher income gain, increased production as reflected by larger herd size among the respondents resulting into their high perception of benefit through internet source.

Place of access to Internet: The correlation if any between the place of access to Internet (kiosks or home) and perceived benefits of all respondents were dealt by employing rank order correlation. All the respondents as well as their perceived benefits were combined and depicted in Table 6.37 to arrive at the following results.

The calculated value of rank order correlation (R) from the above results was 0.47; this positive correlation indicated that the place of access to Internet had not influenced the farmers' perceived benefits. This is because of the equality among speed, quantity and type of information accessibility through Internet.

Computer literacy: The correlation if any between the computer literacy and perceived benefits of respondents were dealt by employing pearson's correlation coefficient and subjected to t-test to found significance level at required degrees of freedom.

Table 6.36: Coefficient of concordance ('w' value) between different categories of livestock farmers and perception of benefits rendered by Internet

S. No.	Statements(s)	Small (=44)		Medium (n=162)		Large (n=74)	
		Total score	Rank order	Total score	Rank order	Total score	Rank order
1.	Internet can be used to meet the information needs of the community	3	6.5	9	10.0	4	9.0
2.	Commercial and economic development opportunities are made effective through Internet communication	3	6.5	11	9.0	5	8.5
3.	Internet communication has converted the world into global village and kept the global market under one umbrella	4	5.0	19	4.0	5	8.5
4.	Internet has made access to high speed communication	6	3.5	24	1.0	6	7.0
5.	Internet communication has certainly overruled the middlemenship	9	1.0	22	2.0	8	4.5
6.	Employment opportunities are made possible through Internet	6	3.5	20	3.0	9	3.0
7.	Consumer oriented farming has made possible through Internet	8	2.0	13	7.0	7	6.0
8.	Internet communication has created demand for poultry products	2	8.5	12	8.0	8	4.5
9.	Internet communication has made convergence of animal husbandry technologies	2	8.5	15	6.0	12	1.0
10.	Internet communication makes an individual conversant	1	9.0	17	5.0	10	2.0

Table 6.37: Rank order correlation coefficient ('R' value) between Place of access and prerceived benefits

S. No.	*Statements(s)*	*Kiosks (n=234)*		*Home (n=46)*		*d*	*d^2*
		Total score	*Rank order*	*Total score*	*Rank order*	*Total score*	*Rank order*
1.	Internet can be used to meet the information needs of the community	20	7.0	3	7.5	-0.5	0.25
2.	Commercial and economic development opportunities are made effective through Internet communication	27	4.0	1	10.0	-6.0	36.00
3.	Internet communication has converted the world into global village and kept the global market under one umbrella	14	10.0	4	5.5	4.5	20.25
4.	Internet has made access to high speed communication	29	3.0	6	4.0	-1.0	1.00
5.	Internet communication has certainly overruled the middlemenship	31	2.0	2	9.0	-7.0	49.00
6.	Employment opportunities are made possible through Internet	32	1.0	7	3.0	-2.0	4.00
7.	Consumer oriented farming has made possible through Internet	25	5.0	8	2.0	3.0	9.00
8.	Internet communication has created demand for poultry products	16	9.0	4	5.5	3.5	12.25
9.	Internet communication has made convergence of animal husbandry technologies	21	6.0	9	1.0	5.0	25.00
10.	Internet communication makes an individual conversant	19	8.0	3	7.5	0.5	0.25

'R'=0.47

Table 6.38: Correlation between computer literacy and perceived benefits

S.No.	*Mean*		*Variance*		*'r' value*
	Computer literacy	*Perceived benefit*	*Computer literacy*	*Perceived benefit*	
1.	12.42	14.11	6.01	6.54	0.3473**

** = $p<0.01$

The coefficient of correlation (r=0.3473), which is positive and highly significant ($p<0.01$) indicates that the literacy level in computer was associated with perceived benefits, thus it is clear that the respondents with higher computer literacy had perceived more benefits through Internet. This may be due to higher intensity of Internet access, bilingual/multilingual accessibility of websites reflected by higher computer literacy. If the owners are unaware or do not understand the information and communication technologies available, they are unlikely to grab the benefits rendered by them.

Level of information intensity: The correlation if any between the level of information intensity and perceived benefits of respondents were dealt by employing pearson's correlation coefficient and subjected to t-test to found significance level at required degrees of freedom.

Table 6.39: Correlation between perceived benefits and information intensity.

S.No.	*Mean*		*Variance*		*'r' value*
	Computer literacy	*Perceived benefit*	*Computer literacy*	*Perceived benefit*	
1.	12.86	9.81	9.98	9.73	0.4496**

** = $p<0.01$

From the positive and highly significant ($p<0.01$) coefficient of correlation value ($r=0.4496$), it could be concluded that the level of information intensity was associated with perceived benefits. Level of information intensity influences the farmers to avail or not to avail Internet services. Large amounts of data and many transactions are likely to motivate the respondents to access the Internet services that can help to attain various tangible benefits offered by Internet.

Problems Faced During Information Acquisition and Suggestions Given by the Livestock Farmers

This section of the chapter was presented and discussed in the following two sub-heads:

Problems Faced by the Livestock Farmers During Information Acquisition

The livestock farmers had experienced some problems while acquiring information through Internet; they were noted and presented in Table 6.40.

Table 6.40: Distribution of respondents as per the problems faced during information acquisition.

S.No.	*Problems faced*	*Frequency*	*Percentage*
1.	Domination of English language	137	48.93
2.	Interrupted power supply	81	28.93
3.	Illiteracy in Internet access	99	35.36
4.	No separate cabins for women at CICs	19	6.79
5.	Distant location of CICs	109	38.93
6.	Connectivity	92	32.86
7.	Complexity of the information available through Internet	63	22.50
8.	High cost	117	41.79
9.	No access to verbal conversation	70	25.00

From Table 6.40, it was evident that majority (48.93%) of the respondents faced the problem of domination of English language, followed by high cost (41.79%), distant location of CICs (38.93%), illiteracy in Internet access (35.36%), connectivity (32.86%), interrupted power supply (28.93%). No access to verbal conversation with experts (25.00%), while 22.50 per cent of the respondents faced the problem of complexity of the information available through Internet. Only 6.79 per cent of the women respondents had felt that there were no separate cabins for them at CICs.

Suggestions Given by the Livestock Farmers to Overcome the Problems During Information Acquisition

The livestock farmers who availed Internet services for information acquisition were asked to give suggestions to overcome the problems/constraints faced by them during the process of information acquisition and the same are presented in Table 6.41.

It is clear from Table 6.41 that majority (51.43%) of the respondents suggested that concerned SMSs must be online, in order to clarify their FAQs followed by 45.00 per cent of the respondents who suggested to go for pictorial presentation of the information, which may facilitates the farmers to acquire more knowledge and influences the improvement of effectiveness of Internet communication.

Majority of the livestock farmers (43.21%) suggested to establish CICs at all village head quarters. As envisaged in Kuppam project to open CICs in all villages, to enable the educated rural people to have direct access to latest information in various aspects.

Similarly, 41.79 per cent of the respondents suggested to provide all livestock farming related websites in local languages. Apart from CDAC, "Anusaraka" an ongoing project based in Hyderabad, which has Sanskrit grammar and computational

Table 6.41: Distribution of respondents according to their suggestions

S.No.	*Suggestions*	*Frequency*	*Percentage*
1.	All livestock farming related websites must be in simple and local language	117	41.79
2.	Supply of battery operated hardware	81	28.93
3.	An initial orientation program to new comers of CICs	99	35.36
4.	Separate cabins for women at CICs	19	6.79
5.	Pictorial presentation of the information	126	45.00
6.	To clarify the FAQs, Subject Matter Specialists must be online	144	51.43
7.	Complete information about livestock farming must be available under one website or package	107	38.21
8.	Establishment of CICs at all village head quarters	121	43.21
9.	Borrowing dishnet connection	101	36.07

linguistics to develop machine translation systems between English and other languages and among Indian languages.

A percentage of 38.21 of the respondents sought complete information about livestock farming under one website and 35.36 per cent of the respondents suggested to arrange initial orientation program to new comers. As envisaged in future, ICTs projects and forthcoming CICs need to be established in rural areas for the development of Internet communication.

More than one-third of the respondents (36.07%) suggested to borrow Dishnet connection to CICs, which may be used to down load online publications and other services like question and answer sessions on specialized subjects could be possible.

A percentage of 28.93 of respondents suggested to provide with battery operated hardware and 6.79 per cent of the women respondents sought separate cabins at CICs, which may greatly aid to their knowledge and in turn motivate them for increased frequency of visits to kiosks to avail Internet services.

SUMMARY AND CONCLUSION 7

The salient findings of the study, conclusions, implication drawn thereupon and suggestions for future research have been presented in this chapter.

Summary

India is a country in which agriculture occupyies a central position in the life of the people and their socio-economic aspects. Despite the growth of the other aspects of the economy, agricultural sector still provides a livelihood for approximately 67 per cent of the population, dwelling in rural areas and provides employment to about 60 per cent of the country's labour force. Moreover, revenue of this vital sector continues to provide nearly 18 per cent of the total GDP of the country. In addition to this, livestock proved to be an integral part of livelihood system of more than 80 per cent of rural households who maintain one or other type of livestock as a source of income and risk absorbing factor in case of crop failure. The share of livestock sector is 6.30 per cent of the total GDP of the country and is deeply affected by global economic, environmental and marketing forces.

Without communication capabilities, required to access, analyse and share the information the livestock sector remains at mercy of global market forces. Technological discoveries of 1990s like radio, television and wide range of audio-visual aids have already broughtout considerable changes and made profound impact on rural population and at present the world is entering into an information age and developments in ICTs such as telecommunication, teleconferencing and Internet are offering new and cost effective way to reach the disadvantaged population in particular. Among which Internet is emerging as a tool of improved communication and information access, directly related to social and economic development of rural population and helped to reach the un-reached.

The ongoing IT revolution has opened up huge opportunities for providing valuable information and this was in fact executed through various pilot projects, viz. Warna wired project in Kolhapur and Sangli districts and Bhabaleswar project in Ahmednagar district of Maharashtra, Info village in Pondicherry and Gyandoot project in Madhya Pradesh. All pilot projects and conclusively demonstrated that, if the information is available at the door step, farmers are willing to pay and make use of technical, marketing and general information. In spite of number of activities and projects oriented towards the transfer of ICTs, still the acceptability and utilization of ICTs were found to be very low in the rural arena. This created the need to probe knowledge, utility, practicality and constraints in utilization of ICTs by livestock farmers.

On the similar lines of above said initiatives, the government of Andhra Pradesh has started Kuppam project three year ago in Chittoor district, wherein the Information and Communication Technology (ICT) tools were extensively used to disseminate outreach information to the endusers at the cutting edge. So far no study attempted to evaluate the effectiveness of these initiatives. The study at hand is aimed to provide certain information, viz. how does IT work? which section of the farming community has benefited most? what kind of information do farmers at Internet seek? what information is actually available on Internet? what is the attitude of the farmers towards Internet? how do the farmers exploit the advantages of Internet? what is the feasibility of Internet communication in the backward areas where sizeable population is below the poverty line? These questions require definite answers before expanding Internet communication in India.

Keeping all these points in view the study entitled "Internet Communication among Livestock farmers in Chittoor district of Andhra Pradesh" is conceptualised with an general objective to study and analyse the factors that are determining the effectiveness of internet communication with special reference to livestock development in Kuppam area of Chittoor district in Andhra Pradesh and the specific objectives are as follows:

1. To study the livelihood profile and Information seeking behaviour of livestock farmers.
2. To assess the kind of Information available through Internet as perceived by the livestock farmers.
3. To measure the effectiveness of Internet communication in terms of knowledge and perceived benefits among livestock farmers.
4. To study the factors that influence the perceived benefits of the livestock farmers.

Methodology

Following *ex post facto* research design, the present study was carried out in Chittoor district of Andhra Pradesh, which was selected purposively. Kuppam, V. Kota, Shanthipuram, Ramkuppam and Gudipalli mandals in the district were selected for the study, as these mandals were considered to be the world's first i-community and only mandals in introducing Chief Minister's Empowerment Yojana (CMEY) program to setup a chain of CICs at mandal headquarters. Thus all the villages under the jurisdiction of these five mandals were included in the study and by following proportionate random sampling method, 80 respondents from Kuppam, 80 from V.Kota, 60 from Shanthipuram, 40 from

Ramkuppam and 20 respondents from Gudipalli, were selected to constitute the sample size of 280.

Salient Findings of the Study

1. More than seventy per cent of the livestock farmers were highly literate and 55 per cent of their families' educational status was medium. Majority (59.69%) of the respondents possessed medium communication aids. Majority (64.64%) of the Internet users belongs to medium extension contact category and most of the respondents (63.93%) had shown medium inquisitiveness towards the Internet communication.
2. Most of the livestock farmers were landless and 77.50 per cent of the respondents earned their income through livestock. Majority (59.50, 50.58, 57.50 and 54.25 per cent) of cattle, buffalo, sheep and poultry owners, respectively possessed medium sized herd. Majority (42.98, 45.21, 72.50 and 73.91 per cent) of the respondents followed intensive, range, semi-intensive and deep-litter rearing system in cattle, buffalo, sheep and poultry, respectively. Most of the cattle and buffalo owners fed both roughages and concentrates, while most of the sheep owners gone for only grazing and majority of the poultry owners fed starters, growers and finishers to their stock.
3. Majority (38.57%) of the respondents were using the Internet for the last 1-2 years. At kiosks majority of the respondents (29.30%) visited the Internet once in a week, while at home majority (8.60%) of the respondents availed the Internet services daily.
4. Majority (50.85 and 39.13 per cent) of the respondents preferred to avail Internet services during afternoon hours and highest (46.58 and 60.87 per cent) of the

respondents spent time at Internet about 1 to 2 hours at kiosks and home, respectively.

5. Most of the respondents (44.64%) covered less than 2 km distance to reach the kiosks. While a major junk (57.50%) preferred Telugu language to access the information through Internet and majority 46.40 per cent of the respondents had intermediate computer literacy.
6. The livestock farmers had visited at a total of thirteen websites, i.e. three bilingual and eight anilingual. From among the bilingual sites www.aponline.com ranked first, while among the anlingual sites www.poultrysolutions,com ranked first based on their regular visits. From among the available websites, livestock farmers could able to access 65.00, 71.43 and 72.50 per cent of technical, marketing and general information, respectively.
7. Among technical information, majority (55.98%) of livestock farmers obtained the information to full extent pertains to feeding, while majority (43.42%) of livestock farmers revealed that the information about fodder production was very useful. Whereas, in market information, majority (57.45%) of livestock farmers obtained the information to full extent about current price of livestock products and 47.96 per cent of livestock farmers revealed that the information about marketing of livestock products was very useful. In general information, majority (55.55%) of the livestock farmers obtained the information pertaining to rural finance to full extent, while 47.69 per cent of the respondents revealed that the information pertains to education was very useful.
8. 80.16, 42.46, 32.15 and 89.13 per cent of cattle, buffalo, sheep and poultry owners, respectively perceived that the

technical information was available through Internet. While 80.99, 68.49, 30.00 and 86.96 per cent of cattle, buffalo, sheep and poultry owners, respectively perceived that the marketing information was available through Internet. Whereas 76.03, 72.60, 52.50 and 80.43 per cent of cattle, buffalo, sheep and poultry owners, respectively perceived that the general information was available through Internet.

9. 60 per cent livestock owners had changed their habit of information acquisition after being acquainted with Internet.
10. Majority (51.31%) of the livestock farmers considered the Internet source as trustworthy, while 49.79, 45.70, 43.37 and 33.19 per cent of the livestock farmers agreed that the Internet source is free from error, latest, complete and informs to a greater degree about a subject, respectively.
11. All the species owners possessed very fair knowledge pertaining to technical, marketing and general aspects of livestock production. Among which, 27.15, 15.83, 9.48 and 40.74 per cent of the knowledge score was obtained through Internet source among cattle, buffalo, sheep and poultry owners, respectively.
12. A considerable number of livestock farmers obtained different aspects of technical knowledge exclusively through Internet. Majority (7.50%) of the respondents obtained knowledge about various conventional feeds followed by different methods of breeding (6.43%), while 3.93 per cent of the respondents gained knowledge about different feed additives. Only 3.21 per cent of the respondents had attained knowledge pertains to diet manipulation to change egg-yolk colour.
13. Majority (97.42% and 82.50%) of the cattle and buffalo and sheep owners perceived that Internet has made

access to high-speed communication. Among poultry owners majority (93.48%) perceived that Internet communication has created demand for poultry products and ranked first.

14. From among the sheep owners' majority of them (82.50%) had perceived that Internet has made access to high-speed communication, while 42.50 per cent of the sheep farmers had perceived that, employment opportunities are made possible through Internet communication.
15. Among all the species owners, the independent variables socio-economic status, herd size, total income, language, extension contact, inquisitiveness, frequency of use, utility of information and credibility were found to be positively and significantly correlated with knowledge level of livestock farmers. Whereas, age was found to be negatively and significantly correlated with knowledge and gender, type of information, time of use and distance covered were found to be non significant.
16. Among all the species owners, the independent variables socio-economic status, herd size, total income, language, extension contact, inquisitiveness, frequency of use, utility of information and credibility were found to be positively and significantly correlated with perceived benefits of livestock farmers. Whereas age, gender, type of information, time of use and distance covered were found to be non-significant.
17. The knowledge level of three categories (small, medium and large) of livestock farmers differ with each other and are in complete agreement with the perception of benefits rendered by Internet.
18. The knowledge level of livestock farmers differ, based on the place of access. But had not influenced the farmers' perceived benefits.

19. The positive and highly significant ($p<0.01$) value of coefficient of correlation indicates that the literacy level in computer and level of information intensity were highly associated with perceived benefits.
20. Majority (48.93%) of the respondents faced the problem of domination of English language, followed by costly technology (41.79%), distant location of CICs (38.93%), illiteracy in Internet access (35.36%), connectivity (32.86%), interrupted power supply (28.93%), no access to verbal conversation with experts (25.00%), while 22.50 per cent of the respondents had faced the problem of complexity of the information available through Internet. Only 6.79 per cent of the women respondents had felt that, there were no separate cabins for them at CICs.
21. Majority (51.43%) of the respondents had suggested that concerned SMSs must be online followed by pictorial presentation of the information (45.00%), establish CICs at all village head quarters (43.21%), provision of all livestock farming related websites in local languages (41.79%), complete information about livestock farming under one website (38.21%), initial orientation program to new comers (35.36%), borrow Dishnet connection to CICs (36.07%), provide with battery operated hardware (28.93%) and only 6.79 per cent of the women respondents sought separate cabins at CICs.

Conclusion

A fair proportion of young as well as women respondents availed the Internet services irrespective of place of access of information. Respondents came from higher educational status and with high level of inquisitiveness availed more Internet services and preferred Telugu language to access the same.

Respondents belong to large category possessed more knowledge and the knowledge acquisition through Internet sources among all species owners was very fair. The Internet communication has changed the information acquisition habits among the respondents to a greater extent. Respondents with higher literacy level in computer applications and belong to large category perceived more benefits in comparison to their counter parts.

Irrespective of gender, time of use and distance covered respondents availed the Internet services besides skilled in computer applications and utilized the technical information most followed by general and marketing information. Extension contact, language, inquisitiveness and utility of information were found to be highly effective in improving knowledge and the benefits attained through Internet communication.

Domination of English language, high cost, distant location of CICs and lack of knowledge in Internet were the major problems/ constraints perceived by all the respondents, irrespective of literacy levels and economic conditions.

Cyber Livestock Communication model conceptualized to explain the process of Internet communication of livestock information in the countryside was found to be applicable and to describe it factually.

Implications of the Study

It is hoped that the present investigation will come out with the information that would assist the planners, organizers, administrators, research scholars, IT officials, NGOs and other agencies which are involved in promoting the Internet communication in the rural arena. From the results of the study, the following implications were drawn:

1. High level literacy among the livestock farmers has made the governmental authorities and other agencies to launch various ICTs for agricultural and rural development, which was reflected in the study. So similar projects can be undertaken in high level literacy areas in order to facilitate the livestock farmers to obtain more benefits. Since literacy of the livestock farmers is playing a crucial role in utilizing the services of ICTs, concern authorities should concentrate on improving the literacy rate of the livestock farmers.
2. Though there is significant role of Internet communication, women are not given enough importance. Therefore, the planners, organizers and IT officials should look at development in relation to equity for women. The most important factor contributes towards the social change is the level of information disseminated among the rural farm women, which would be achieved effectively through Internet communication.
3. The study will help the IT officials, government authorities, NGOs, service providers and other communication agencies in improving the credibility and effectiveness of Internet communication in rural and remote areas.
4. Internet may further widen the gap between developed and developing countries, between rich and poor farmers and between those who know how to make use of the new technologies and those who do not, i.e. popularly known as digital divide. Thus an appropriate pro-poor government policies and strong grassroots programmes are necessary to change the direction and spread the benefits of Internet equitably among the rural poor and disadvantaged groups of the country.

5. Low levels of inquisitiveness among livestock farmers about new ICTs can be improved by making the information locally relevant, practically and economically viable in unserved or under served areas and the technologies nust be designed in such a way that a totally unprepared poor person can immediately get the hang of it without feeling threatened.
6. Destabilization that is rooted in food insecurity, poverty and low literacy level can be improved through efficient communication. Improved access to information among local people and their ability to produce and share information, referred as best informed citizens would be less destabilized.

In support of the above implications, the measures suggested at administrative, financial and institutional levels could actually serve the purpose of bringing both rich and poor closer and bridge the existing gaps. On the other hand, some of these measures could directly or indirectly contribute towards strengthening Internet communication process in rural and remote areas. Establishing CICs at all village headquarters seems to be more essential measure in the short run.

Suggestions for Future Research

The results of the present study revealed the need for future research in general directions as suggested below:

1. The present investigation was confined to a particular area in Andhra Pradesh. The study needs to be replicated on larger samples covering larger area, so that the inferences drawn can be generalized to a greater extent and explore various theoretical explanations.

2. There is an acute need to study whether the information and communication technologies have broughtout the much talked 'Gender equality'.
3. Need assessment of livestock farmers pertains to the availability of information through Internet may be investigated.
4. A comparative study among the attitude of livestock farmers towards Internet communication and other communication technologies may prove worthwhile.
5. Another study could be devoted towards the content analysis of information about veterinary and animal husbandry technologies available through Internet sources will certainly pave the way.

APPENDICES

APPENDIX - I

Mandal-wise Population and Literacy Percentage

S.No.	Name of the Mandal	Population (in numbers)	Literacy percentage
1.	Kuppam	91,000	44
2.	V.Kota	63,000	34
3.	Shanthipuram	53,000	26
4.	Ramkuppam	41,000	21
5.	Gudipalli	27,000	14

APPENDIX - II

Cattle Equivalents for Different Categories of Livestock

S.No.	Category		Cattle equivalent
I.	**Cattle**		
1.	Cattle in milk	Indigenous	1.00
		Crossbred	1.19
2.	Dry cattle	Indigenous	0.80
		Crossbred	0.85
3.	Cattle heifer	Indigenous	0.65
		Crossbred	0.85
4.	Young stock	Indigenous	0.40
		Crossbred	0.48
5.	Bulls and Oxen		1.26
II.	**Buffalo**		
1.	Buffalo in milk		1.26
2.	Dry buffalo		1.01
3.	Buffalo heifer		0.69
4.	Young stock		0.40
5.	Buffalo bulls and working stock		1.26
III.	**Sheep/Goat**		0.10
IV.	**Poultry (100 adult poultry)** 0.005		

APPENDIX - III

Dr. H.P.S. ARYA
Head & Principal Scientist
Division of Extension Education

Indian Veterinary Research Institute
Izatnagar - 243 122 (U.P.)

Dated..........................

Dear Sir/Madam,

I wish to inform you that one of my Ph.D. student Dr. G.R.K. Sharma has taken up a research study entitled "Internet Communication among Livestock Farmers in Chittoor District of Andhra Pradesh" as part of his doctoral degree programme. In this connection he has devised a judgement sheet consisting of various components to measure personal, livelihood and Internet characteristics of livestock farmers which forms the independent variables.

I am happy to inform you that you are selected as one of the judges and your valuable responses in the enclosed judgement sheet will contribute significantly for the research. I request you to spare your valuable time in going through the responses to the items listed in the sheet. Since we are behind the schedule of time for completing the research project, I request your kind cooperation in the early response of your judgement.

I thank you in advance for your kind cooperation and valuable response.

With kind regards,

Yours sincerely,

[H.P.S. ARYA]

APPENDIX - IV

Judgement Sheet for Identifing the Independent Variables

Please find here with some of the components which indicate personal, livelihood and Internet characteristics of livestock farmers. Please indicate the degree of relevance of each component which is suitable for measuring the knowledge level and perceived benefits of livestock farmers. In a 3-point continuum, if you perceive any component has more optness then please put a (v) mark corresponding to such component in the column of 'more relevant', follow the same procedure in case of relevant and less relevant columns.

S.No.	*Component*	*More relevant*	*Rele-vant*	*Less relevant*	*'Z'*
1.	Age				0.5159*
2.	Gender				0.5212*
3.	Socio-economic status				0.5792*
4.	Physical facilities				0.4142
5.	Herd size				0.8446*
6.	Economic motivation				0.3137
7.	Total income				0.6004*
8.	Language				0.5995*
9.	Information storage pattern				0.2632
10.	Information processing pattern				0.3554
11.	Frequency of use				0.5478*
12.	Time of use				0.5669*
13.	Channels of communication				-1.6744
14.	Distance covered				0.6466*
15.	Rural orientation				-0.7321
16.	Extension contact				0.5221*
17.	Participativeness				0.3311
18.	Credibility				0.8020*
19.	Coordination				0.4468
20.	Kind of information				0.5992*
21.	Utility of information				0.6462*
22.	Attitude towards Internet				0.2007
23.	Inquisitiveness				0.6112*
24.	Achievement motivation				0.3366

* Variables selected for the study

APPENDIX - IV

Division of Veterinary Extension Education
Indian Veterinary Research Institute
Izatnagar - 243 122 (U.P.)

INTERVIEW SCHEDULE

Respondent No.___________
Date of Interview ___________

Section - I

(I) Personal, socio-economic and livelihood Profile of Livestock farmers

1. Name of the respondent : ______________________
2. Name of the village : ______________________
3. Name of the Mandal : ______________________
4. Name of the C.I.C. : ______________________
5. Age (in completed years) : ______________________
6. Sex : male (2) Female (1)
7. Caste : SC & ST(1) BC(2) OBC(3) GEN(4)
8. Religion : Hindu(3) Muslim(2) Christian(1)
9. Education of livestock farmers and their family members:

Members		*Illiterate*	*Functionally Literate*	*Primary School*	*Middle School*	*High School*	*Inter-mediate*	*Graduate & above*
		0	1	2	3	4	5	6
Self								
Family memb-ers in num-bers	M							
	F							
	T							

M = male, F = female and T = total

10. Marital status:
    i. Married (2)
    ii. Unmarried (1)

11. Family type:
    - i. Nuclear (2)
    - ii. Joint (1)
12. Family size:
    - i. Small (1-3 numbers) (1)
    - ii. Medium (4-6 numbers) (2)
    - iii. Large (7-9 numbers) (3)
    - iv. Very large (above 9 numbers) (4)
13. Material possession:
    - Furniture (1)
    - Bullock cart (2)
    - Farm implements (3)
    - Pump set (4)
    - Tractor (5)
    - Bicycle (1)
    - Mobike (4)
    - Car (5)
14. Possession of communication aids:
    - Radio set (1)
    - Television set (2)
    - V.C.P. (3)
    - V.C.R. (3)
    - Tape recorder (1)
    - Telephone (4)
    - Computer and Internet (5)
15. Type of house:
    - i. Thatched shed (1)
    - ii. Mud walls and thatched (2)
    - iii. Brick walls and titled (3)
    - iv. Concrete house (4)
    - v. Concrete double house (5)
16. Position in the family:
    - i. Head (3)
    - ii. Independent (2)
    - iii. Dependent (1)

17. Do you consider livestock farming as a main occupation?

    Yes ❑ No ❑

    If 'No' please indicate your main occupation

    ____

18. Are all the family members engaged in the same occupation?

Yes ❑ No ❑

If 'No' please indicate various other occupations, where your family members are engaged:

a. ..

b. ..

c. ..

(II) Land holding

i. Landless (0)
ii. Marginal (below 1 ha.) (1)
iii. Small (1.1 to 2 ha.) (2)
iv. Low medium (2.1 to 4 ha.) (3)
v. High medium (4.1 to 10 ha.) (4)
vi. Large (above 10 ha.) (5)

(III) Herd Size

1. Please indicate the number of cattle/buffalo owned (female):

		Number(s)			
Species	Breed	Milch	Dry	Heifers	Calves
Cow	Crossbred	❑	❑	❑	❑
	Desi	❑	❑	❑	❑
Buffalo	Crossbred	❑	❑	❑	❑
	Desi	❑	❑	❑	❑
	Total	❑	❑	❑	❑

2. Please indicate the number of cattle/buffalo owned (male):

		Number(s)			
Species	Breed	Draft	Bulls	Young bulls	Calves
Cow	Crossbred	❑	❑	❑	❑
	Desi	❑	❑	❑	❑
Buffalo	Crossbred	❑	❑	❑	❑
	Desi	❑	❑	❑	❑
	Total	❑	❑	❑	❑

3. Please indicate the number of sheep/goat owned:

Species	Number(s) Male	Female	Young
Sheep	❑	❑	❑
Goat	❑	❑	❑
Total	❑	❑	❑

4. Please indicate the number of poultry owned:

- Egg type :
- Meat type :

(IV) Total Annual Family Income

i. From agriculture ____________
ii. From dairying ____________
iii. From sheep farming ____________
iv. From goat farming ____________
v. From poultry ____________
vi. Service ____________
vii. Farm labour ____________
viii. Other sources (If any) ____________

Total Income: ....

(V) Livestock Rearing Practices and Production

A. Cattle

i. Milk yield per day (litres) .

Species	*Time* *Morning*	*Evening*	*Total*	*Price per litre*
Cow				
Buffalo				
Total				

ii. Quantity of milk consumed : _________ in litres/day

iii. Quantity of milk sold/day :

- Sale of liquid milk : ______ in litres to farmers/society cooperative
- Sale of products : ______ locally/near by town

iv. Average amount received through sale of male calves: Rs._____ per annum

v. Average amount received through sale of dung: Rs._____ per annum

vi. Rearing system : Range (1) Semi-confinement (2) Confinement (3)

vii. Number of grazing hours :__________hrs.

viii. Feeding: Only roughages (1)
Only concentrates (2)
Both (3)

B. Sheep/Goat

i. Amount received through sale of animals : Rs.___________

ii. Amount received through sale of skin : Rs.___________

iii. Amount received through sale of manure : Rs.___________

C. Poultry

i. Rearing system: Deep litter (1) Cage (2)

ii. Feeding system: Starter & Grower (1) Starter, Grower & Finisher (2)

iii. Amount received through sale of eggs Rs.___________

iv. Amount received through sale of birds Rs.___________

v. Amount received through sale of deep litter Rs.___________

Section - II

(I) Are you a subscriber of Newspaper/Magazine/Journal related to livestock farming? Yes (2) No (1)
If 'Yes' please mention in detail.

1.
2.
3.

(II) Extension Contact

Please indicate your degree of awareness with the following personnel?

S.No.	*Extension worker*	*Are you aware*	*Weekly once*	*Fort-nightly once*	*Monthly once*	*Once in six months*	*Never*
1.	Gram sevak		4	3	2	1	0
2.	V. D. O.		4	3	2	1	0
3.	V. L. I.		4	3	2	1	0
4.	V. A. S.		4	3	2	1	0
5.	University Personnel		4	3	2	1	0
6.	Agricultural Officer		4	3	2	1	0
7.	C. A. O. Officer		4	3	2	1	0
8.	G. L. O.		4	3	2	1	0
9.	C. R. O.		4	3	2	1	0

(III) Information Seeking Behaviour

What are the sources through which you gain technical and marketing information and please indicate your regularity of contact:

	Sources	*Frequency*			
		R	*O*	*S*	*N*
A.	*Personal - Localite channels*				
i.	Discussion with family members	3	2	1	0
ii.	Discussion with friends/relatives	3	2	1	0
iii.	Discussion with neighbours/fellow farmers	3	2	1	0
iv.	Discussion with progressive farmers	3	2	1	0
v.	Others, if any	3	2	1	0
B.	Personal-Cosmopolite Channel				
i.	Discussion with officials of state A.H. department	3	2	1	0
ii.	Discussion with officials of state agricultural University.	3	2	1	0

Contd...

iii	Discussion with officials of Devp. Depts.	3	2	1	0
iv	Discussion with representatives of private input agencies	3	2	1	0
v	Discussion with progressive farmers from other villages	3	2	1	0
vi	Other if any	3	2	1	0
C	Impersonal - Cosmopolite Channel				
i	Listening to farm radio broadcast	3	2	1	0
ii	Reading from magazines	3	2	1	0
iii	Viewing farm telecast	3	2	1	0
iv	Accessing through Internet	3	2	1	0
v	Visiting to CICs	3	2	1	0
vi	Field visits/Pashu melas/Kisan melas	3	2	1	0
vii	Others if any	3	2	1	0

R= regularly, O=often, S=seldom, N=never

(IV) Internet profile of livestock farmers

i. Are you visiting Internet kiosks Yes (2) No (1)

ii. Is the information available at Internet kiosks helpful in day-to-day life? Yes (2) No (1)

iii. Where do you access Internet and at what interval?

Place	*Frequency*			
	Once in month	*Once in a week*	*2-3 times in week*	*Daily*
At home	1	2	3	4
At kiosk	1	2	3	4

iv. How much time you do spend at the Internet kiosks?

Time	*Preference*	*Duration*		
		(0 - 1) hrs.	*(1 - 2) hrs.*	*(2 - 3) hrs.*
Morning	❑	1	2	3
Afternoon	❑	1	2	3
Evening	❑	1	2	3

v. Distance covered by you (If assessed from kiosks)

a. Less than 2 km (1)
b. 2 - 5 km (2)
c. 6 - 10 km (3)
d. 10 - 15 km (4)

vi. Please indicate to what extent you know the following languages?

Languages	*Do not Know*	*Understand*	*Understand & Speak*	*Understand Speak & Read*	*Understand Speak Read & Write*
English	0	1	2	3	4
Hindi	0	1	2	3	4
Telugu	0	1	2	3	4
Other (s)	0	1	2	3	4

vii. Does the usage level of English language, influence the access of the Internet? Yes (2) No (1)

viii. In which language you prefer to access the information through Internet?

a. English (1) c. Telugu (3)
b. Hindi (2) d. Other(s) (4)

ix. Please indicate your expertness in using the following technologies?

Technology	*Beginner*	*Intermediate*	*Advanced*
Computer	1	2	3
Internet	1	2	3
E-mail	1	2	3

x. What kind of information is available for the farmers on Internet?

S.No.	*Category*	*Yes*	*No*	*Specific Information*
A.	*Technical information*			
1.	Breeding	2	1	
2.	Feeding	2	1	
3.	Management	2	1	
4.	Health care	2	1	
5.	Vaccination schedule	2	1	
6.	Fodder production	2	1	
7.	Preparation of milk products	2	1	
8.	Project proposal writing	2	1	
B.	Marketing information			
1.	Price list of medicines or vaccines	2	1	
2.	Sale & purchase of livestock	2	1	
3.	Marketing of livestock products	2	1	
4.	Current price of livestock products	2	1	
5.	Status of National and International market	2	1	
C.	General information			
1.	Rural finance	2	1	
2.	Insurance facilities	2	1	
3.	Education	2	1	
4.	Employment	2	1	
5.	Mandi gyan	2	1	
6.	Weather	2	1	
7.	Governmental schemes	2	1	

xi. Kindly indicate how different Internet sites are visited by you for receiving Information and at what frequency:

	Regularity		
Name of the site	*Regular*	*Often*	*Seldom*
	3	2	1
	3	2	1
	3	2	1

xiii. Kindly indicate the extent of information obtained and usefulness of information available through Internet

Categary	*Extent of information obtained*			*Extent of usefulness of information*			
	Full information	*Some information*	*Little information*	*Very useful*	*Useful*	*Somewhat useful*	*Very little useful*
Breeding							
Feeding							
Management							
Health control							
Vaccination schedule							
Fodder products							
Milk products making							
Marketing/Auction of livestock							
Marketing of milk/meat/egg							
Insurance facilities							
Loan facilities							
Rural finance							
Education							
Employment							
Weather							
Central/state government schemes							

xii. Please indicate the problems/constraints faced by you regarding information acquisition:

A
B
C
D
E

xiii. Please offer your suggestions to overcome them and for better acquisition of information through Internet:

A
B
C
D
E

SECTION - III

(I) Credibility:

A set of statements is given below which represents credibility of Internet sources of information. Please state your degree of agreement with each statement in the following 5 point scale:

S.No.	*Statement(s)*	*Degree of agreement*				
		SA	*A*	*UD*	*DA*	*SDA*
1.	Information available through Internet is trustworthy in nature	5	4	3	2	1
2.	Information available through Internet is free from error and correct in nature	5	4	3	2	1
3.	Information available through Internet is complete and adequate in nature	5	4	3	2	1
4.	The Internet sources informs to a greater degree about a particular topic	5	4	3	2	1
5.	The Internet sources maintains and provides latest and timely information	5	4	3	2	1

SA=strongly agree, A=agree, UD=undecided, DA=disagree, SDA=strongly disagree

(IV) Perceived Benefits

Please indicate the benefits (as perceived by you) rendered by Internet as main source of communication:

1.
2.
3.
4.
5.

KNOWLEDGE TEST - I

Following are some of the questions representing Cattle and buffalo farming. Please give answers according to your knowledge and also indicate how you have acquired the information:

S.No.	*Questions*	*Correct*	*Incorrect*	*If 'correct'* *Through Internet*	*Through Other Sources*
A.	*Breeding*				
i.	Enumerate the indigenous breeds of cattle	❑	❑	❑	❑
ii.	Enumerate the exotic breeds of cattle	❑	❑	❑	❑
iii.	What is the correct time of insemination?	❑	❑	❑	❑
iv.	After how many days, does the cow repeats heat cycle?	❑	❑	❑	❑
v.	Does the animal becomes pregnant without fail through A.I.?	❑	❑	❑	❑
vi.	Enumerate the methods of breeding of cattle	❑	❑	❑	❑
B.	*Feeding*				
i.	What do you mean by balanced feed?	❑	❑	❑	❑
ii.	What are the feed supplements added to cattle feed?	❑	❑	❑	❑

iii.	Enumerate the unconventional feed sources of cattle?	❑	❑	❑	❑
iv.	What is maintenance ration?	❑	❑	❑	❑
v.	Do you feed extra ration to pregnant animals, If so why?	❑	❑	❑	❑
vi.	Do you feed extra ration to lactating animals, If so why?	❑	❑	❑	❑
vii.	What is the dry matter requirement per 100 kg body weight?	❑	❑	❑	❑
viii.	What are advantages of urea treated paddy straw?	❑	❑	❑	❑
ix.	What are the advantages of urea-molasses-liquid-diet?	❑	❑	❑	❑
x.	Enumerate the different systems of grazing?	❑	❑	❑	❑
C.	*Management*				
i.	What is the right time of feeding colostrum?	❑	❑	❑	❑
ii.	How many days the baby calf should receive colostrum?	❑	❑	❑	❑
iii.	What is the appropriate age of weaning of calves?	❑	❑	❑	❑
iv.	When do the calves are dehorned?	❑	❑	❑	❑
v.	When do the male calves are castrated?	❑	❑	❑	❑
vi.	What steps you follow for clean milk production?	❑	❑	❑	❑
vii.	What criteria should be considered in cattle housing?	❑	❑	❑	❑
viii.	Enumerate the systems of housing in cattle	❑	❑	❑	❑
D.	*Health care*				
i.	Why do you isolate the sick animals?	❑	❑	❑	❑
ii.	What is the correct time of deworming of young ones?	❑	❑	❑	❑

iii.	What control measures you have taken against external parasites in cattle?	❑	❑	❑	❑
iv.	How do you diagnose the common diseases at farm level?	❑	❑	❑	❑
v.	What first-aid measures you follow for common ailments?	❑	❑	❑	❑
E.	*Vaccination*				
i.	Indicate the vaccination schedule followed in cattle?	❑	❑	❑	❑
ii.	What is the appropriate time for H.S. vaccination?	❑	❑	❑	❑
iii.	What is the appropriate time for FMD vaccination?	❑	❑	❑	❑
iv.	What measures should be taken before vaccination?	❑	❑	❑	❑
F.	*Fodder production*				
i.	Enumerate the important fodder crops	❑	❑	❑	❑
ii.	What is the correct time of sowing the fodder seeds?	❑	❑	❑	❑
iii.	How many times irrigation to be done in fodder fields?	❑	❑	❑	❑
iv.	At what stage the first cut is done?	❑	❑	❑	❑
v.	How many cuts you can expect per year?	❑	❑	❑	❑
vi.	What forage crops are used for silage making?	❑	❑	❑	❑
vii.	What forage crops are used for hay making?	❑	❑	❑	❑
viii.	What procedure you follow for making good silage?	❑	❑	❑	❑
ix.	What procedure you follow for making good hay?	❑	❑	❑	❑
x.	After how many days, silage is ready for feeding?	❑	❑	❑	❑

G	*Preparation of milk products**				
i.	What are the milk products available?	❑	❑	❑	❑
ii.	Ingredients required for preparation of sweets?	❑	❑	❑	❑
iii.	Making procedures of various milk products	❑	❑	❑	❑
H.	*Project proposal writing**				
i.	Points to be considered in writing a Project proposal?	❑	❑	❑	❑
ii.	Assets to be possessed before submission?	❑	❑	❑	❑
I.	Marketing*	❑	❑	❑	❑
i.	Price list of medicines/vaccines	❑	❑	❑	❑
ii.	Sale and purchase of livestock	❑	❑	❑	❑
iii.	Marketing of livestock products	❑	❑	❑	❑
iv.	Current price of livestock products	❑	❑	❑	❑
v.	Status of National and International market	❑	❑	❑	❑
J.	*General information**				
i.	Rural finance	❑	❑	❑	❑
ii.	Insurance facilities	❑	❑	❑	❑
iii.	Education	❑	❑	❑	❑
iv.	Employment	❑	❑	❑	❑
v.	Mandi gyan	❑	❑	❑	❑
vi.	Weather	❑	❑	❑	❑
vii.	Governmental schemes	❑	❑	❑	❑

* Indicates the modification made to be original scale

KNOWLEDGE TEST - II

Following are some of the questions representing Sheep farming. Please give answers according to your knowledge and also indicate how you have acquired the information:

S.No.	*Questions*	*Correct*	*Incorrect*	*If 'correct'* *Through Internet*	*Through Other Sources*
A.	*Breeding*				
i.	Enumerate the indigenous breeds of sheep	❑	❑	❑	❑
ii.	Enumerate the exotic breeds of sheep	❑	❑	❑	❑
iii.	Enumerate methods of breeding of sheep	❑	❑	❑	❑
iv.	Indicate the breeding seasons in sheep	❑	❑	❑	❑
B.	*Feeding*				
i.	What do you mean by balanced feed?	❑	❑	❑	❑
ii.	Enumerate the unconventional feed sources of sheep?	❑	❑	❑	❑
iii.	What is maintenance ration?				
iv.	Do you feed extra ration to pregnant animals, if so why?	❑	❑	❑	❑
v.	Enumerate different grazing systems?				
C.	*Management*				
i.	What is the right time of feeding colostrum?	❑	❑	❑	❑
ii.	How many days the lamb should receive colostrums?	❑	❑	❑	❑
iii.	What is the appropriate age of weaning lambs?	❑	❑	❑	❑
iv.	When do the male lambs are castrated?	❑	❑	❑	❑
v.	What criteria should be considered in sheep housing?	❑	❑	❑	❑
vi.	Enumerate the housing systems in sheep?	❑	❑	❑	❑
D.	Health care				

i.	Why do you isolate the sick animals?	❑	❑	❑	❑
ii.	What is the correct time of deworming?	❑	❑	❑	❑
iii.	At what intervals deworming is performed?	❑	❑	❑	❑
iv.	What do you understand by dipping?	❑	❑	❑	❑
v.	How do you diagnose the common diseases at farm level?	❑	❑	❑	❑
vi.	What first-aid measures you follow for common ailments?	❑	❑	❑	❑
E.	*Vaccination*				
i.	Indicate the vaccination schedule followed in sheep?	❑	❑	❑	❑
ii.	What is the appropriate time for E.T. vaccination?	❑	❑	❑	❑
iii.	What is the appropriate time for FMD vaccination?	❑	❑	❑	❑
iv.	What measures one should take before vaccination?	❑	❑	❑	❑
F.	*Project proposal writing**				
i.	Points to be considered in writing a Project proposal?	❑	❑	❑	❑
ii.	Assets to be possessed before submission?	❑	❑	❑	❑
G	*Marketing**				
i.	Price list of medicines/vaccines	❑	❑	❑	❑
ii.	Sale and purchase of livestock	❑	❑	❑	❑
iii.	Marketing of livestock products	❑	❑	❑	❑
iv.	Current price of livestock products	❑	❑	❑	❑
v.	Status of National and International Market	❑	❑	❑	❑
H.	*General information**				
i.	Rural finance	❑	❑	❑	❑
ii.	Insurance facilities	❑	❑	❑	❑
iii.	Education	❑	❑	❑	❑
iv.	Employment	❑	❑	❑	❑
v.	Mandi gyan	❑	❑	❑	❑
vi.	Weather	❑	❑	❑	❑
vii.	Governmental schemes	❑	❑	❑	❑

* Indicates the modification made to be original scale

KNOWLEDGE TEST - III

Following are some of the questions representing poultry farming. Please give answers according to your knowledge and also indicate how you have acquired the information:

S.No.	*Questions*	*Correct*	*Incorrect*	*If 'correct'* Through Internet	*If 'correct'* Through Other Sources
A.	*Breeding*				
i.	How do you select individual males/ females for breeding purpose?	❑	❑	❑	❑
ii.	Enumerate the methods of mating in poultry	❑	❑	❑	❑
iii.	What do you understand by moulting?	❑	❑	❑	❑
iv.	How does moulting is related to bird's laying quality?	❑	❑	❑	❑
B.	*Feeding*				
i.	Why do poultry require more concentrates?	❑	❑	❑	❑
ii.	What do you mean by feed additives?	❑	❑	❑	❑
iii.	Enumerate the feed additives that promote feed intake	❑	❑	❑	❑
iv.	Enumerate the feed additives that facilitate digestion	❑	❑	❑	❑
v.	Enumerate the feed additives that affect health status	❑	❑	❑	❑
vi.	Indicate the methods of feeding systems of poultry	❑	❑	❑	❑
vii.	What is a starter ration?	❑	❑	❑	❑
viii.	What is a finisher ration?	❑	❑	❑	❑
C.	*Management*				
i.	At what age layer starts laying eggs?	❑	❑	❑	❑

ii.	What is the appropriate marketing age and weight for broilers?	❑	❑	❑	❑
iii.	What are the criteria's to be considered in poultry housing?	❑	❑	❑	❑
iv.	Enumerate the systems of housing in poultry	❑	❑	❑	❑
v.	Indicate the advantages of deep litter system	❑	❑	❑	❑
vi.	What is the correct age of debeaking?	❑	❑	❑	❑
vii.	Debeaking is performed to which beak?	❑	❑	❑	❑
viii.	How does the debeaking is done?	❑	❑	❑	❑
ix.	What do you understand by artificial lighting management?	❑	❑	❑	❑
D.	*Health care*				
i.	When should deworming in birds be done?	❑	❑	❑	❑
ii.	What control measures you follow against external parasites in birds?	❑	❑	❑	❑
iii.	What drugs to be added in feed?	❑	❑	❑	❑
E.	*Vaccination*				
i.	What is the correct age for R.D.?	❑	❑	❑	❑
ii.	What is the correct route for R.D. vaccination?	❑	❑	❑	❑
iii.	What is the immunity period for single dose of R.D. vaccine?	❑	❑	❑	❑
iv.	What is the correct age for fowl pox vaccination?	❑	❑	❑	❑
v.	What is the correct route for fowl pox vaccination?	❑	❑	❑	❑
vi.	What is the immunity period for single dose of fowl pox vaccine?	❑	❑	❑	❑

E.	*Project proposal writing**				
i.	Points to be considered in writing a Project proposal?	❑	❑	❑	❑
ii.	Assets to be possessed before submission?	❑	❑	❑	❑
F.	*Marketing**				
i.	Price list of medicines/vaccines	❑	❑	❑	❑
ii.	Sale and purchase of livestock	❑	❑	❑	❑
iii.	Marketing of livestock products	❑	❑	❑	❑
iv.	Current price of livestock products	❑	❑	❑	❑
v.	Status of National and International market	❑	❑	❑	❑
G	*General information**				
i.	Rural finance	❑	❑	❑	❑
ii.	Insurance facilities	❑	❑	❑	❑
iii.	Education	❑	❑	❑	❑
iv.	Employment	❑	❑	❑	❑
v.	Mandi gyan	❑	❑	❑	❑
vi.	Weather	❑	❑	❑	❑
vii.	Government schemes	❑	❑	❑	❑

* *Indicates the modifications made to the original scale.*

BIBLIOGRAPHY

Agricultural Statistics (2001) Ministry of Agriculture. Government of India. New Delhi.

Aronson, E., Turner, J. and Carismith, J.M. (1963) Comunication credibility and communication discrepancy as determinants of opinion change. *Journal Abnormal Sociology and Psychology*. Vol. 67: 31-68.

Arunabha Ghosh and Nandan Kamath (2001) Is the internet really the leveller? Information for the civil society. India Together. (arunabha.ghosh@balliol.ox.ac.uk and nandankamath@hotmail.com).

Ayaz, M. (1989) Role of mass media in disseminating agricultural information. *Journal of Rural Development and Administration*. Vol. 22 (1): 108-112.

Babb, S.F. (1998) The Internet as a tool for creating economic opportunity for individuals and families. Ph.D. dissertation, University of California. Los Angeles.

Balit, S., Rios, M.C. and Masies, L. (1996) *Communication for Development for Latin America. A Regional Experience.* FAO, Rome, Italy.

Bhaskaran, C. and Sethu Rao (1985) Correlates of interpersonal communication behaviour of farmers in differentially progressive villages. *Indian Journal of Extension Education*. Vol. 21 (3 & 4): 16-19.

Bhogle, S.K. (1971) Farm information: A case study. *Kurukshetra.* May issue. pp. 15.

Bogue, D.A. (1962) Some tentative recommendations for sociological correct family planning communication and motivation programme in India. *Indian Journal of Extension Education*. Vol. 15(1 & 2): 75-76.

Cecchini, S. and Scott, C. (2003) Can information and communication technology applications contribute to poverty reduction? Lessons from rural India. (e-mail to first author: scecchini@eclac.cl).

Chambers, R. and Guijt, I. (1995) Understanding farmers' communication networks: An experience in the Philippines. FAO report. Rome.

Chambers, R. and Guijt, I. (1996) PRA - five years later: where are we now? World Wide Web publication of the Forest, Trees and People Network of the International Development Research Centre, the Swedish University of Agricultural Sciences, Uppsala, Sweden and the FAO Community Forestry Unit.

Chandel, S.R.S. (2002) *A hand book of agricultural statistics*. Achal Prakashan Mandir. Kanpur.

Charan, A. and Sharma, S. (1989) Socio-economic profile of milk farmers: A study. *Kurukshetra*. 37: 11-16.

Chi-cheng Chang (2001) A study on evaluation and effectiveness analysis of web-based learning portfolio. *British Journal of Educational Technology*. Vol. 32 (4) : 435-458.

Choole, R.R. and Rahudkar, W.B. (1978) Source credibility pattern among small and big farmers. *Indian Journal of Extension Education*. Vol. 24 (1 & 2): 34-39.

Das, P. (2002) 'New Horizons and Extension Modalities'. Memorial lecture delivered on 28.9.02 at ANGRAU, Hyderabad.

Down to Earth (2001) Reducing poverty through agricultural sector: Strategies in eastern and southern Africa. *Popular Media.* Issue (12). pp: 302.

Down to Earth (2001) The information must wired. *Popular Media.* Issue (12). pp: 407.

Edwards, P.N. and Wajcman, J. (1995) *Feminists theories of technology. Handbook of Science and Technology Studies.* Sage Publications. pp. 257-285.

Ernest, R.S. (1987) A study of communication utilization behaviour small and big farmers and its implication to communication strategy. Unpublished Thesis. Indian Agricultural Research Institute. New Delhi.

FAO. (1995) Farmer-first approaches to communication: a case study from the Philippines. Rome.

FAO. (1996) Communication for development in Latin America: a regional experience. GCP/RLA/114/ITA. Unpublished project report. Rome.

Flanigan, A. and Metzger, Majority (2000) Perceptions of Internet Information Credibility. *Journalism and Mass Communication Quarterly.* 77(3): 515-540.

Goswami, A. (1996) Development of cognitive learning scale to test the knowledge of sheep farmers about different managemental practices. *Indian Journal of Animal Health.* Vol. 35(1) : 17-21.

Greenberg, J.H. (1966) *Lying: Moral Choice in Public and Private Life,* New York, pp. 32.

Grimes, S. (2000) Rural areas in the information society : diminishing distance or increasing learning capacity? *Journal of Rural Studies.* Vol. 16(1) : 13-21.

Heap, Nick, Ray, Robin Mason, Hughie Mackay (1995) *IT and society. A readers.* Saga, London.

Heinich, R., Molenda, M., Russell, J., & Smaldino, S. (1996). *Instructional Media and Technologies for Learning.* Columbus: Prentice Hall.

Hiltz, S. and Johnson, K. (1999) User satisfaction with computer-mediated communication systems. *Management Science*. 36(6): 739-764.

Horgan, John (June 1992). "Claude E. Shannon: Unicyclist, Juggler, and Father of Information Theory." IEEE Information Theory Society Newsletter. (April 1992). "Claude E. Shannon." *IEEE Spectrum*. pp. 72-75.

Hudson, H. (1995) *Economic and social benefits of rural telecommunication.* FAO, Rome, Italy.

Jung, K. (2001) Internet connectedness and Inequality. *Journal of Computer-Mediated Communication*. 28(4:) 507-553.

Kenesten, K. (2001) Grassroots of ICT projects in India. Prelimnary Hypothesis. E-mail attachment from kken@mit.edu (requested from the author via mail).

Keringert, N. F. (1983) *Foundations of Behavioural Research.* Surjit Publications. New Delhi.

Khaldoun Almhanna, M.D. (1998) Accessing the quality of health information on the Internet. Health on the Net foundation. Available online at www.hon.com

Kim, Deok (2002) Internet gender gap in Asia. *The Korean Times*. http://www.hankooki.com

King, J.L. and Martin, M.J. (1999) *The saga of cyberspace construction boomtown in the great divide. Culture of the Internet.* San Diego, Academic Press. pp 3-33.

Kirkman, G. (1999) Its more than just being connected: A discussion of some issues of Informaiton Technology. Paper presented at the Development of E-commerce workshop. Massachusetts Institute of Technology. Aug'16-17. (www.cid.harvard.edu/ciditg/resources.pdf).

Kofi, Annan (1997) Global Knowledge. Address to World Bank Conference. UN-Secretary General.

Krueger, A. (1995) How computers have changed the wage structure : Evidence Microdata from 1984 to 1992. *Quarterly Journal of Economics*. Vol. 108(1) : 33-61.

Kulkarni, P.S. (1973) 'A study of socio-economic factors affecting the adoption of Artificial Insemination in cattle in selected villages of key village center, Nagpur'. Unpublished Thesis, Jawaharlal Nehru Krishi Vishwa Vidyalaya, Jabalpur.

Lishan, Adam (1996) Electronic communications technology and development of Internet in Africa. *Information Technology for Development*. Vol. 7 : 133-144.

Maniar, A. (2002) A study on Internet usage for educational purpose by the students of the Maharaja Sayajirao University of Baroda. Vadodara.

Manjunath, L. and Marilingannavar, R.N. (1989) Socio-economic characteristics and existing livestock farming practices in Dharwad district of Karnataka. *Karnataka Journal of Agricultural Sciences*. 5(10): 59-62.

Manjunatha, C. (1991) A study on the interests, knowledge and attitude of young workers on self-employment in Ranga Reddy District of Andhra Pradesh. *Dairymen*. 32(1&2): 64-68.

Meier-Dallach H.P. (1998) *Communication, Organization and Territory: Understanding Information, Business, Technology and Geography.* London. pp: 145-164.

MSSRF (2000) Internet reaches rural poor. *Appropriate Technology*. Vol. 27 (1) : 8-11.

Munyna, H. (2000) 'Information and communication technologies for rural development and food security; farmers from field experiences in developing countries'. Workshop on the role of information and communication technologies in rural development and food security. November, 2000, SD Dimensions, FAO, Rome, Italy.

Murthy, R.K. (1990) Factors associated with knowledge of livestock farmers. *Maharastra Journal of Extension Education.* 9(1): 181-184.

Narmatha, N., Subramanian, R. and Mathiyagan, P. (1995) Knowledge level of farm women in scientific poultry farming. *Poultry Advisor.* Vol. 28(4) : 53-56.

Negroponte, N. (1995a) *Being digital.* London: Hodder & Stoughton. Chapter 6.

Negroponte, N. (1995b) 'Survey of rural economic development officers'. Unpublished paper presented to the Ontario Agriculture Electronic Communications Working Group, Canada, January 1995.

Newcomb, T.M. (1953) *Theories of Communication: An Introduction,* Cambridge University, Cambridge, Eng., pp. 124.

Nie, Norman, H. and Lutz, E. (2001) *Internet and Society: A preliminary report. The Digital Divide.* Cambridge. MIT Press. pp. 269-274.

Pigato, M. (2001) Information and Communication Technology, Poverty and Development in Sub-Saharan Africa and South Asia. World Bank. Washington, D.C. (www.worldbank.org/afr/wps).

Pye (1958) 'Communication behavior of small farmers and its influence on knowledge, symbolic and use adoption'. Unpublished Ph.D. thesis, Div. of Agril. Extension, IARI, New Delhi.

Ramanand, P.R. (2001) The Internet in Thailand: towards a culture of responsibility. *Media Asia.* Vol. 28 (4) : 183-205.

Ramchand, Chakravarthy, Rand Mahipal (1990) Dairy innovations, factors affecting adoption by farmers. *Milcow.* 12: 25-26.

Rao, B.S. and Kherde, R.L. (1990) A standardized test to measure the knowledge of dairy farmers livestock production technologies. *Asian Journal of Dairy Research.* 9:224-228.

Rao, M.R. (1982) Impact of cooperative societies on their beneficiaries in Guntur district of Andhra Pradesh. *Journal of APAU Research. 23* (1): 39-41. Reach. www.economist.com/languagesonthenet/countries).

Richardson, D. (1995) The Internet and rural development: Recommendations for strategy and activity. Final report, FAO. Rome. Italy.

Richardson, D. (1996) The Internet and rural development: Recommendations for strategy and activity. Paper published online at http://www.fao.org/wiacent/faoinfo/sustdev/cddirect/contents.html/)

Richardson, D. (1997) The Internet in rural and agricultural development: An integrated approach. FAO, Rome, Italy. pp. 77.

Richardson, D. and Paisley, L. (1998) The first mile of connectivity. FAO. Rome. (www.fao.org/docrep.html).

Rios, M.C. (1999) Farmer—first approaches to communication—A case study from Philippines. Communication for Development. Rome.

Rogers, E. M. (1995) *A History of Communication Study: A Biographical Approach.* New York: The Free Press.

Rogers, E.M. and Sharma, S.K. (1971) A communication research approach to communication the diffusion of innovation in diffusion research needs. Missouri Agricultural Experiment Station. *North Central Regional Research Bulletin. Columbia.* pp. 2-26.

Ruth, S.R. (2000) Measuring long term effects of technology transfer in developing nations : the case of Internet training at the Romanian Academy of Science. *Information Technology for Development*. Vol. 9 : 105-121.

Samantha, K. (1993) Functional analysis of livestock farming. *Indian Journal of Animal Production and Management*. Vol. 2 (1): 69-73.

Sandhu, A. (1973) Relative efficiency of farm methods of measuring credibility of farm information sources. *Indian Journal of Extension Education*. Vol. 9 (1 & 2): 71-74.

Sawhney, M.M. (1967) Farm practices—adoption and the use of information sources and media in rural community in India. *Rural Sociology*. Vol.32 (3): 310-323.

Schweiger, W. (2000) Media credibility: Experience or image. A survey on the credibility of world wide web in comparision to other media. *European Journal of Communication*. 15(1): 37-59.

Sharma, V.P. (1999) *Cyber Extension: Potential advantages*. MANAGE. Hyderabad.

Shashaani, Lily (1997) Gender differences in computer attitudes and use among college students. *Journal of Educational Computing Research*. Vol. 16 (1) : 37-51.

Sherman, R.C., End, C. and Gardner, J. (1998) Sex differences in Internet use : Attitudinal and Experimential Correlates. Presented to the American Psychological Society Convention, Washington.

Singhal, A. and Brown, W.J. (2000) The entertainment-education-communication strategy : Past Struggles, Present Status and Future Agenda. *Journal Komunikasi*. Vol. 12 : 19-36.

Singhal, A. and Rogers, E.M. (2001) *Rising technopolicies. India's Communication Revolution: From bullock cart to cyber mart*. Sage Publications. Wageningen Agricultural University. The Netherlands.

Snedor, G.W. and Cochron, W.G. (1994) *Statistical Methods.* Iwoa State University Press. United States of America.

Sood, A.D. (2000) *A social investor's Guide of ICTs for development. The communication initiative.* Center for Knowledge Societies. December 3. Bangalore.

Sundaraswamy, B. (1993) A study on extent of knowledge and profits gained through livestock farming. *Karnataka Journal of Agricultural Sciences*. 6(2): 187-189.

Swaminathan, M.S. (1993) *Information Technology: Reaching the Unreached.* Chennai. India.

Tallero, E. and Gaudette, P. (1995) *Harnessing information for development : A proposal for World Bank.* Washington.

Tiamiyu, M.A. (2000) Availability, accessibility and use of information technologies in Nigerian federal agencies: A preliminary survey. *Information Technology for Development.* Vol. 9 : 91-104.

Turban, E.; Mclean, E. and Wetherbe, J. (2000) *Information Technology for management : Making connections for strategic advantage.* John Weilly & Sons Inc., New York.

Tyagi, K.C. and Sohal, T.S. (1984) Effectiveness of film shows in Kisan Vikas Kendra villages. *Annual Report.* National Dairy Research Institute. Karnal.

Uhegbu, A.N. (2000) Information communication networking in rural communities: The case of Imo state, Nigeria. *Journal of Information Science.* Vol. 26 (1) : 51-59.

Venkataramaiah, P. (1983) Revised in (1990) 'Development of socio-economic scale'. Ph.D. Thesis, Andhra Pradesh Agricultural University. Hyderabad.

West, M.D. (2000) Defining and measuring credibility of internet sources: Developing a scale. *Journalism and Mass communication Quarterly. 66:* 277-284.

World Bank (1995) Harnessing information for development: a proposal for a World Bank group vision and strategy by E. Talero and P. Gaudette.

Yadav, K.P.S. (2001) Virtually there. Centre for Science and Environment. *Down to Earth.* 9(18): 89-96. (www.cseindia.org/publications.htm)

Zijp Willem (1994) 'Improving the transfer and use of agricultural information: A guide to information technology.' Washington, D.C. World Bank.

INDEX